JOURNEY
AMONG ANIMALS

JOURNEY AMONG ANIMALS

THE ANIMAL STORIES AND MEMOIRS OF A ZOOLOGIST

MEREDITH HAPPOLD

The Book Guild Ltd

First published in Great Britain in 2021 by
The Book Guild Ltd
9 Priory Business Park
Wistow Road, Kibworth
Leicestershire, LE8 0RX
Freephone: 0800 999 2982
www.bookguild.co.uk
Email: info@bookguild.co.uk
Twitter: @bookguild

Typeset in Aldine401 BT

Printed and bound by CPI Group (UK) Ltd, Croydon, CR0 4YY

ISBN 978 1913551 919

British Library Cataloguing in Publication Data.
A catalogue record for this book is available from the British Library.

For David, with my infinite love and gratitude.

CONTENTS

(ANIMAL STORIES LISTED IN ITALICS)

ACKNOWLEDGEMENTS

Everyone mentioned in this book, and many others, have been a part of my life, and words cannot express the depth of my gratitude to them. Above all, I am grateful to my husband, David, who has been my dearest and best friend and colleague for more than fifty years, and to my wonderful family. All of the animals mentioned in this book have also been a part of my life and I am most grateful to them too, for their friendship and trust. They have taught me so much and I owe it to them to tell their stories.

The research mentioned in this book was supported by Monash University and La Trobe University in Victoria, Australia, the University of Ibadan in Nigeria, the Australian National University in the A.C.T., Australia, and Chancellor College, University of Malawi, Zomba, Malawi, and permission to carry out fieldwork in Malawi was granted by the Malawi Department of National Parks and Wildlife and the Department of Forestry. Also, the research could not have been carried out without access to specimens, and the assistance of the staff, of many museums. My deepest thanks to all these people and institutions, and to everyone else who helped us with our studies.

The photographs were taken by David or me unless stated otherwise in the captions, and I am very grateful to David for scanning the original slides and negatives.

I am also very grateful to Audrey Young and David Happold for reading the manuscript and for their very valuable comments, and to Judy Colville,

Clive Crouch, Prue Gill, Robert Golding, Ian Harvie, Paula Jenkins, Elisabeth Larsen, Tony Lee, Kathleen McCracken, Isabel MacPherson, Anita and Elias Sherz, Keiko Schmeisser, Michael and Nadine Watson, Helen Wegman and Pat Woolley for their encouragement and useful comments on sections of the manuscript.

PROLOGUE

Why am I setting out, at the age of three score and ten, to write an account of my life as a zoologist? There are several reasons. I have been lucky enough to work on wild mammals in Australia, Nigeria and Malawi and have had many adventures while doing so. The study of animals has changed enormously in my lifetime, so I think it is worth recording my career for its historical value. Also, I have taught many students who wanted to become zoologists and am aware that many people of all ages love animals and enjoy reading about them. But, above all, I want to describe some of the wonderful encounters that I have had with special animals because these individuals have taught me so much about themselves and I owe it to them to share what I have gained from their trust.

These animals include the goats, dogs, horses and other animals that were part of my childhood, the native Australian hopping-mice and other native mice whose social behaviour I studied for my Ph. D., and the mongooses, genets, hyraxes, mole-rats and other animals that my husband and I kept as pets and/or for our research while we were living in Nigeria and Malawi. They also include the elephants and other wild animals encountered during fieldwork in Africa, the possums and birds in our gardens in Australia, and the bats I have studied – particularly in Malawi.

Bats get such bad press! In our culture, they are associated with bad things – ghosts, darkness, getting tangled in people's hair to mention only three. In Africa they are feared as evil spirits. All this is so unfair! Bats are warm, intelligent, potentially friendly and very delightful animals and I want to tell stories that will surely change attitudes towards them. I want people to know why I have loved so many of them. The Chinese believe that bats bring good luck and they have certainly brought good luck to me. My husband and I have been able to publish many scientific papers about various aspects of their natural history, and I became the editor and one of the authors of the profiles of bats for a definitive, six-volume work, *Mammals of Africa,* which was published in 2013. The search for information about African bats has taken me to fascinating places with wonderful people.

So yes. I want to share my experiences and I owe it to all the animals who have shared their lives with me to tell their stories.

1

SETTING OUT WITH A GOAT AND DREAMING OF HOPPING-MICE

Because *Journey among Animals* is largely about the delightful experiences I have had with animals during my youth and subsequent career as a zoologist, let me begin with little Mickey.

Mickey, who made me love bats

Mickey was not a mouse, but a mouse-eared bat – a Large-footed Myotis. He was a little, very dark grey bat with a sweet face, tiny eyes and alertly pricked ears, who was captured somewhere in tropical Australia by two German bat biologists who were spending a sabbatical year at Monash University where I was a student. They were Professor Franz Möhres who was studying how insectivorous bats use echolocation to perceive their surroundings, and Dr. Erwin Kulzer who was investigating how bats regulate their body temperatures, and other aspects of their physiology. Mickey was one of Erwin's study animals. He was fed with hand-held offerings of mealworms, and he was given drops of water from the tip of Erwin's finger. At the end of his year at Monash, Erwin, who loved his bats, gave some of them to me.

I took the bats home. Three of them were of the same species – Little Northern Free-tailed Bats. They were nervous little creatures that always kept together, and they were far more content with their own company than with mine. But, from the very beginning, Mickey was different. He was full of curiosity about humans, and as willing as the nicest of dogs to be friendly and affectionate. Following Erwin's advice, I continued to feed him on mealworms and other insects that I held in my fingers, and I let him lick drops of water from my fingertips. I kept him in my bedroom and let him fly around the room during the night. Bats are not instinctively afraid of humans and so it was not surprising that he sometimes landed on me. Whenever possible, I gave him a mealworm when he did this, and inevitably he came more and more often. One evening, before feeding him, I decided to try to train him to come when I called him. To do this, I began to flick a finger and thumb together to make a loud click whenever I anticipated that he was going to land on me. These clicks contain ultrasonic sounds as well as the noise that we can hear and therefore they are easily heard by insectivorous bats whose ears are very sensitive to ultrasonic sounds. In effect, these clicks became calls to bring the bat to me – and he learned to come when he was called within one hour! And, once Mickey realized that a click meant that he was being called, he would always come immediately whether or not he was hungry, whether or not he was rewarded with food, and whether or not it was me who was calling him!

At night, I slept in a bed that was placed alongside my large bedroom window, and one of the curtains hung quite close to my pillow. Mickey was free to fly around my bedroom at night, but how often he did so I do not know. However, he always had a flight at dawn soon after I woke up, and it became his habit to end the flight by landing on the curtain, climbing down it, dropping onto my pillow and scuttling under my hand, which I cupped over the pillow to form a tiny cave. Mickey could fit very snuggly under my cupped hand, and he was delightfully warm and furry. I often went back to sleep after he had tucked himself under my hand and, by the time I woke up again he was always sound asleep.

Mickey always had water in the cage in which he slept during the day, and I assume he went into his cage if he was thirsty in the night. But most of his meals were of mealworms that he took from my fingers. He used to eat three and then, if he were in his cage, he would scuttle to the water and drink. However, I often fed him when he was outside his cage – perhaps clinging to me or perhaps resting on my desk. When this happened, he would eat three mealworms and then indicate that he was thirsty by licking the tip of my finger. Because both Erwin and I had often given him drops water from our fingers, it was easy to guess the meaning of this signal, and I responded by giving him water. Then, once he had trained me to give him a drink on demand, he started to let me know when he wanted a mealworm. There is a parallel with dogs of course. We train them not to urinate in our houses, but every dog has to train its owner to recognize a particular signal that means "I need to go outside" and every dog has its own way of signalling this message. Mickey asked me for food with a particular posture and buzzing vocalization, and a particular look of expectation. I was surprised by this because I had not expected it. But I was studying animal behaviour at Monash and I decided to see what he would do if I gave him a mealworm when he had asked for water, and vice versa. To my absolute astonishment, he delivered an unmistakably clear signal that meant "No! No!! NO!!!" He uttered a loud buzzing vocalization and banged the elbows of his wings several times on whatever he was resting on, and I had the distinct impression that he was furiously chastising me! For several weeks, I used to give him the wrong thing often enough to test the consistency of his response. The really fascinating thing was that he always said "No" when I offered the wrong thing, even though he must surely have been thirsty on some of the occasions I offered water, and hungry on some of the occasions I offered food. Was he actually training me?

I learned from Erwin Kulzer that Mickey would go into hibernation when the weather where I lived started to get cold in autumn and, in anticipation of this, I built him a hibernaculum. This was a small wooden box with a landing stage, a steep ramp leading up to a ceiling of wire mesh that he could easily climb over and eventually hang from, and a series of wooden partitions that gave him a choice of dimly lit places, or a totally dark place, in which to hibernate. He liked his box – especially its darkest end – and we developed a routine. Once a week I would open the box and get him out. At first, he was invariably very cold and barely capable of moving, which is normal during hibernation. I warmed him up in my hands, and he also shivered to generate heat and then, when his body temperature had risen, he would take off, fly around the room and then come to me for food and water. Finally, when he was satiated, he would fly to the landing stage of his box and turn around so that he could 'look' at me by means of his echolocation calls – a very sophisticated form of radar. I could not hear these extremely loud ultrasonic calls, but I could tell when he was echolocating by seeing his open mouth and by hearing very faint audible sounds. I invariably approached and stroked his back, and then he would scuttle backwards up the ramp and disappear into the darkest part of the box.

Then, out of pure kindness, I made a bad mistake. I reasoned that, because he was a social bat, he would hibernate in the wild with others of his kind. They would all be warm at first, and then they would lose body-heat quite slowly as they dropped into hibernation. I decided to simulate this by placing a jar of warm water in his dark box. The jar was wrapped in felt and Mickey loved to go to sleep huddled against it. Giving him the 'hot water bottle' proved to be a grave mistake because it lowered the humidity in the box and his wings dried out and developed holes that never healed. Eventually he could no longer fly, but he never lost the ability to scuttle around and climb.

However, long before this happened, something incredible took place! One day, I went through the usual routine of waking him up, letting him fly for a few minutes, letting him feed and drink and then fly back to the landing stage of his hibernaculum. As usual, he waited for his pat and then scuttled back to his dark place. But this time, to my amazement, he came racing back down the ramp to the landing stage, and then started buzzing furiously and banging his wings (just like a child throwing a tantrum) and signalling "No. No! NO!!" I had forgotten to heat his hot water bottle!

Sadly, most people from European backgrounds, and most Africans too, are taught to fear and loathe bats. In our literature and movies, they are associated with ghosts, darkness, vampires and wickedness and, in at least some parts of Africa, they are thought to be evil spirits. In fact, both insectivorous bats and fruit bats (including flying-foxes) are delightful, warm, furry, curious, friendly and very intelligent animals, and I am sure that they are capable of cognitive thinking. One expects them to be rather like flying mice but instead, they have a mind that is quite different to that of most mammals of equivalent brain size. I introduced Mickey to all my friends and also to most of my parents' friends, including all the 'girls' my mother went to school with. They were scared of him at first but, before long, some were willing to sit on the floor in a circle and take turns making clicks to call him to them. He would always go to them – scuttling over the floor with the speed of a mouse – and then he would climb onto their hands and gaze into their faces with his lips slightly parted as he 'looked' at them by echolocating. After that, all but one would want to have a turn with him and their ideas about bats were transformed. Of all my friends, there was only one who remained aloof. Mickey captivated everyone else.

Having Mickey was an extraordinary and enlightening experience, but it was not the first and not the last, and I should go back to the beginning of my journey among animals.

Serendipity

All journeys through life begin somewhere at a particular time, and from there the road goes onwards and there is no turning back. The road we follow has forks and intersections, but things happen to guide the way: we meet special people and experience special events that influence our decisions and shape our destinies. I have encountered special animals as well.

My journey began in 1945 at Ivanhoe in Victoria, Australia, where I was born, and I lived in the suburb of Ivanhoe with my parents and younger brother for the first eight years of my life. My mother was repelled by 'creepy-crawlies', but she was so determined that her children would not share her fears that she encouraged us to love and respect all living things by picking up snails, worms and beetles and putting them into our outstretched hands. She also encouraged us to pat a neighbour's foal and the draught horses who pulled the baker's cart, and she took us to the zoo where we rode elephants! We also visited friends who had dogs or cats and, when I was about six years old, we got a dog of our own – a much-loved Cocker Spaniel named Ben. So, animals came into my life very early on and, by the time I was three years old, I had decided to be a 'zoo doctor'. But that changed when Uncle Max came back from Heard Island in 1948.

Uncle Max – Maxwell Crichton Downes – was a very close friend of my parents. In 1947-48, he spent a year on Heard Island – a mountainous and heavily glaciated island at latitude 53°S near the Antarctic. There he studied penguins, petrels, terns, seals and other wildlife, and he fell down a crevasse in a glacier and lost his camera! All this was very exciting! And then, when he came back to Australia, he joined the Victorian Fisheries and Game Department and began to study ducks, bustards and other birds, and I concluded that his life was just perfect. Uncle Max was a zoologist, so I decided to be a zoologist too – and I never changed my mind.

When I was five years old, I went to Ivanhoe State School. There, I got to know the second person who played a major role in determining my destiny. This was Miss Alice McDonald who ran the combined third-grade assemblies, which were held at the beginning of every day, and she was an inspirational enthusiast for nature. She had a table set aside for things of interest such as mushrooms and toadstools, pinecones and seed pods, other plants, bones, tadpoles in jars, and feathers! Miss McDonald was especially interested in birds, and we were encouraged to join the Gould League of Bird Lovers, which was founded in 1909 to honour the work of John and Elizabeth Gould who wrote *The Birds of Australia*, a seven-volume work containing 600 plates (hand-coloured lithographs) that are justifiably famous. Under the influence of Miss McDonald and the Gould League, I might well have become an ornithologist, but one crucial event changed that.

We lived next door to a retired medical doctor who often sat outside on his verandah, and my brother and I enjoyed visiting him and chatting. I often talked to him about birds and I also told him that, now I was in the third grade at school, we were allowed to borrow books from the junior section of the Ivanhoe Public Library. One day, Dr Meigher suggested that I should try to borrow a book from the adult section called *What Bird is That?* by Neville Cayley. This was one of Australia's first comprehensive field guides to Australian birds. I obtained permission to borrow this book and a somewhat bemused librarian went to get it for me. She came back to tell me that, unfortunately, *What Bird is That?* was out on loan. But she held in her hand another book that she thought I might like instead! This was Ellis Troughton's *Furred Animals of Australia*. I borrowed it, and that seemingly insignificant event changed the direction of my journey and eventually even led me to my beloved husband.

Furred Animals of Australia, published in 1941, was the first comprehensive field guide to the mammals of Australia and, like *What Bird is That?*, it was also illustrated with colour plates by Neville Cayley. I loved *all* of those illustrations but the ones that stirred me most deeply were those of the hopping-mice, and the Marsupial Jerboa that is now known by its aboriginal name, Kultarr. Hopping-mice are small, dainty, nocturnal animals that live in deserts and they have large ears, large dark eyes, a long tail with a pencil of black hairs at its tip, and very long hindlimbs that enable them to hop like kangaroos. Kultarrs look very similar and furthermore, they both resemble Winky Jerboa who was my favourite character in the book *Snugglepot and Cuddlepie* by May Gibbs. I loved Winky Jerboa and was so inspired by the illustrations of hopping-mice and the Marsupial Jerboa in Ellis Troughton's book that I cut out some grey cloth, stitched it up and stuffed it to make a jerboa for myself. It did not remotely resemble a jerboa or a hopping-mouse, but I adored it.

In 1953, my parents purchased three acres of land with a dam at Heathmont in the country east of Melbourne. They bought this block of land from Dr and Mrs Colquhoun who owned the adjacent property of ten acres named 'Aringa', which included a large garden, several fields and dams, and some areas of natural bushland. They kept several cows and several Corgi dogs and, best of all, they had a herd of Saanen dairy goats. Saanen goats are supposed to be white, but occasionally a biscuit-coloured one is born and when I first saw the Colquhouns' goats, I fell in love with a biscuit-coloured female kid named Aringa Hebe. About two months later, this kid was included among my Christmas presents! We continued to live in Ivanhoe throughout the next year, but we were able to keep Hebe in the vacant block next door and she was treated in much the same way as our dog. I have vivid memories of Hebe coming with us to Edithvale on Port Phillip Bay, where we spent our school holidays with two of my great aunts. They lived near the sea, so we spent most of the time sunbathing on the sand, swimming in the sea and beachcombing for shells, jellyfish, sharks' eggs and anything else that the tides had washed up. Hebe always came with us. She followed us everywhere – even into the sea! She particularly loved to leap onto our backs if we kneeled down, and sometimes we used to make a pyramid with my father kneeling on the sand, my mother on top of him, me on top of my mother, my brother on top of me, and Hebe prancing about on top of my brother!

During the year that followed the arrival of Hebe into my life, we often visited Heathmont where builders had started to build my parents' dream home and, just before Christmas 1954, when I had just turned nine, we moved in and a new stage in my journey began.

2

'WILLOW HILL' WITH GOATS, HORSES, A POSSUM AND A TRAPPED DOG

My parents named our new home 'Willow Hill' because it was on the side of an east-facing hill with a grove of old osier willows near the top. 'Willow Hill' was a long narrow block divided into three paddocks of equal size and the central one had a dam near its top boundary. All of the paddocks had been cleared for grazing many years earlier, but there were old eucalyptus trees, wattles, native pittosporums and a few old radiata pines around the boundaries. Our house was built between the dam and the grove of willows, as high up the hill as was possible and, from almost every window, there was a magnificent view across a valley of orchards, fields and bushland to the beautiful Dandenong Mountains and the more distant range of alps that included Ben Cairn, Juliet and Mount Donna Buang. It was a long, narrow house with one end at ground level and the other high enough to have a garage and workshop underneath.

As the years passed, a very beautiful garden with many exotic deciduous trees was established around the house, part of the back paddock was fenced off as an enclosure for Hebe and her descendants, a duck-yard with a pond was created for my brother's Khaki Campbell ducks, and the other paddocks were used to produce hay and for grazing horses. We had a large vegetable garden and grew a year's supply of beans, tomatoes and pumpkins as well as greens and other vegetables that are not so easy to preserve. We also established an orchard with apples, pears, peaches, nectarines and plums, and also grapefruit and lemon trees and passionfruit vines – an extraordinary combination of both temperate and semi-tropical plants.

Hebe and other goats

Hebe, the biscuit-coloured goat I was given for Christmas, had kids every year. With help from friends, my father built a very nice goat shed with a milking platform and four stalls in which the goats were locked up at night. At first, my father did all the milking, but I soon learned how to do it and, for many years, the milking was one of my daily chores.

While I was still very young, the goats were my closest friends and I often pretended to be one of them. I would follow them as they went out to graze, lie down with them when they slept or chewed their cuds, watched as they came into season and bleated to attract a mate, and I observed them mating, giving birth and caring for their kids. If they bleated, I bleated back and I tried to understand what all their different calls meant. In short, I tried to act like a goat and, even more important, to *think and feel* like a goat. I soon learned that they were 'creatures of habit' who liked doing the same things at the same time every day. For example, they always expected to be milked in the same order – and chaos reigned if we tried to change that order! And it was always one goat who led the others out to graze, and that same goat who decided when they were to lie down and chew cud. I learned all the subtle nuances of posture and expression that enforced the hierarchy, invited butting contests (in which I participated of course) and elicited play by the kids. And I think I understood the simple meanings of most of their bleats although it is difficult to express them in words, except perhaps for "I am here, where are you?" and "It's all right, nothing to worry about". It was obvious to me that each goat had its own individual personality and that they experienced emotions just as we humans did.

I also learned that goats are clever! One of Hebe's offspring, named Eos, learned that, by resting her front hoofs on the side wall of her stall, she could reach over the door, grasp the pad-bolt between her teeth and slide it across so the gate opened. Having let herself out, she would push up the lids of the food bins, which were usually left unfastened at night, and then help herself to oats, bran and chaff. But the Colquhouns had a goat who was even smarter! One night, there was a terrific thunderstorm and it began to rain. By chance, Mrs Colquhoun looked out of a window and saw, to her horror, that there was a light on in her dairy! Thieves? She and Dr Colquhoun raced out and saw a sight they could hardly believe. Most of the goats, instead of being safe in their stalls, were huddled outside in the rain. The Colquhouns knew that one of their goats, just like Eos, had learned how to open the gate of her stall but, on this occasion, she then let every other goat out of its stall, drove them all outside, shut the goat-shed door so they could not get in again, slid open the pad-bolt on the door into the food shed, pushed open the lids of the food bins and helped herself! After that, I never minded when people called me a silly goat! Those goats taught me all the things I needed to know in order to establish empathy and rapport with animals – wild ones and others – and that proved to be one of the most important things I ever learned.

We had wonderful neighbours at 'Willow Hill'. The Colquhouns lived on 'Aringa', at the top of the hill, and their property adjoined our western boundary. Below us, Mr and Mrs Harvie lived on a seven-acre property named 'Netherby', and our back boundary abutted onto the southern end of a long, narrow bushland block that had been bought by Uncle Max and his wife very soon after my parents bought 'Willow Hill'. Uncle Max continued to be inspirational. I have very clear memories of a camping holiday with him on Phillip Island, which is famous for its colonies of Little Penguins and Koalas. Phillip Island had wonderful, sandy surf beaches, and also rocky headlands with rock pools that teemed with marine life in those days. I was introduced to sea anemones, sea snails of all sorts, crabs, sponges, starfish, little fish and amazing brittle stars. Uncle Max was a great teacher and, under his guidance, he made me work out the basic principles of natural selection and evolution during that camping holiday.

Ruffles, the horse who taught me to ride

The Harvies had a daughter and a son who were older than me. Helen was a keen and very competent horse rider and she had two thoroughbred geldings for shows, eventing and hunting, and a mare for breeding who sadly proved barren. Helen also had a smaller gelding named Ruffles, on whom she had learned to ride. I had always longed for a horse, so I was overjoyed when the Harvies offered me a long-term loan of Ruffles, and Helen said she would teach me to ride! Helen – usually mounted on her beautiful thoroughbred named Camidex – took me out on a leading rein and we rode together along quiet dirt roads and through several large areas of bushland. But it was Ruffles who really taught me to ride.

Ruffles was very calm and quiet with me at first but, as soon as my confidence grew, he began to shy at things! I came off but was told to get straight back on because you had to fall off twelve times before you could be considered a rider! After several falls, I learned how to sit a vigorous shy – and after that he never shied again with me. Instead, he began to pig-root with his head high and, when he could no longer throw me off by doing this, he pig-rooted with his head tucked between his front legs. From pig-rooting he progressed to bucking – and he continued to buck out of sheer high spirits long after I had learned how to sit a vigorous buck. If an adult who was an experienced rider mounted him, he would usually throw in a few bucks and this gave him a bad reputation, but I was to learn that it was not having an *adult* on his back that made him buck.

The butcher in Heathmont had two sons, aged forty and forty-two, who had Down syndrome and were severely handicapped. My mother invited them to 'Willow Hill' so the sons could play with the goats and our dog and have an enjoyable outing. However, they saw old Ruffles in the paddock and what they wanted most of all was to ride him! Although they were hefty, well-grown adults, I was sure that they would be safe on Ruffles – provided we could get them on to his back in the first place! In the end, we got Ruffles to stand beside a bank and somehow we got them on – one at a time of course – and then Ruffles moved forward taking one careful step at a time. One of the lads did not understand that his legs should hang down naturally: instead, he held them akimbo and therefore balanced precariously across the saddle like a see-saw. We had to have one of us holding each leg to prevent him sliding off, and this meant there was no one to guide Ruffles. It didn't matter. He carried each rider up and down our long driveway as though they were tiny children. He stopped every time he felt one of them slipping, and then proceeded with one step at a time. He was amazingly sensitive to their needs.

Camidex, the thoroughbred

Helen's thoroughbred, Camidex, was a very different horse and he came to Helen under unusual circumstances. He had been trained to be a racehorse. Obviously, he had been trained by someone who could ride him, but he bucked off most of his jockeys! Consequently, he had to be retired from racing and he was turned out into a paddock somewhere. At that time, Helen had a great reputation as a rider, and she was often asked to ride other people's horses at shows. At the Royal Melbourne Show one year, she was riding a very eye-catching palomino when a drainpipe under the main arena collapsed under the weight of the horse's hoof and, in front of the crowded grandstand, the horse broke its leg and had to be put down! There was a very kind man among the onlookers who assumed that Helen was riding her own horse. He was the owner of many racehorses whose racing days were over, and he said to Helen, "I have a paddock full of horses. Come and take your pick!" Helen took up this very generous offer and went out to look over the horses. There was one who stood out. He had the most elegant looks, carried his head and tail as though he were a proud stallion, and moved with spectacular grace. Helen picked him immediately, but the man said, "No. You can have any horse except that one. He is impossible to ride." Helen was not convinced and, as the man had said she could have her pick, she insisted on having that one. It was Camidex. He came to 'Netherby' and he allowed Helen to ride him!

Because Camie, as we called him, was an ex-racehorse, his racing-day memories were stirred every Saturday when Helen's grandfather, who lived in a cottage next to one of the Harvies' paddocks, used to listen to the racing commentaries on the radio while he was gardening. Every time a race began, Camie started galloping around the paddock at top speed and, when the race ended, he stopped under a grand old gum tree and tried very hard to buck himself out of his skin. He just *loved* bucking! However, during one of these episodes, he trod on a pointed stick that penetrated through the heel of one hoof and into the fetlock. The wound bled profusely and the situation was extremely serious, but the vet suggested that Helen should nurse Camie for a week before deciding whether or not to put him down. And he recovered!

This drama happened before we came to 'Willow Hill' but, when I was about fourteen, Helen went overseas for an extended period and, during this time, Camie trod on another sharp stick that penetrated the same part of the same hoof! This time, the wound was mainly through scar tissue and was much less serious but, nevertheless, Camie needed a lot of nursing, and it fell my lot to do it. That proved very lucky for me because Camie knew I was helping him and we became very close friends. Then Helen came back from overseas, but she was going to return to marry a man in Switzerland and therefore she had to part with Camie and all of her other horses except Ruffles. Camie had a fine reputation as a dressage horse and eventer, and so he was sold to the Oaklands Riding School where many of the Australian Olympic riders were trained. But it was a disaster! He bucked all the riders off and left a trail of broken arms and sore heads behind him. 'Oaklands' sent him back to 'Netherby' and he was turned out to graze in the paddocks. No one wanted him. But then I received a letter from Helen to say that, if I could ride him, I could have him! Not a loan this time, but a gift! Well, it proved easy. Camie and I were already close friends because I had nursed him and furthermore, I realized that he was not a horse who could be dominated by a demanding rider. Instead, he had to be *asked* – but he would try to do anything I *asked* him to do. And when I rode him, we seemed to fuse, mentally and physically, into one superbeing, and so many of our thoughts were shared.

I had a particularly memorable ride on Camie when I was about fifteen years old. It was Easter, and two friends came to stay overnight. Prue came with her horse, Rillah, and Judy was going to ride Ruffles. We slept in our haystack, which was delightfully soft and smelled wonderful. We turned in early and, at first, we could hear the three horses grazing nearby, and then they came over to the fence and, by stretching their necks, they could reach some of the hay to nibble.

I awoke first. I had been awakened by the bark of a fox, which sounded very close, and I peeped over the edge of the haystack to look for it. The waning gibbous moon had risen over the Dandenong Mountains. A mist lay over the horses' paddock and it was white in the moonlight. The fox trotted by, very close to the haystack, and then slipped under the sliprails into the paddock and melted into the mist. I lay awake with excitement for about half an hour and then it was one o'clock and time to wake Prue and Judy.

We had a hot drink from a thermos, and some sandwiches and chocolate. Then we slid down the haystack, gathered up bridles and crept into the paddock to catch the horses. We all spoke in whispers and were very excited as we saddled-up, rode out of 'Willow Hill' and turned towards the mountain five miles away.

We came to crossroads and saw a red light in the trees. It was the tail light of the milkman's cart and his horse nickered to ours. Then we left the last house behind us and rode through a belt of farming country and orchards. We could smell the cows. The horses were fresh and excited, prancing around and blowing clouds of mist in the frosty air. We let them have their heads and they thundered along in a wild, carefree gallop with their shod hoofs striking bright sparks from the road. Ahead were the Dandenongs and above us the moon and the stars. Then the eastern sky began to glow and, at first light, we reached the foot of Mount Dandenong.

From where we were, a track led straight up the mountain. It ran underneath the cables that carried power to television transmission towers on the top. Although we could have ridden the horses up this track, it was too steep to bring them down again safely, so we tied them up and then climbed up a little way on foot. We climbed just high enough to see the dawn breaking over the valley, and then we had breakfast, which we ate with great gusto. But it was so cold! There was frost on the rocks and grasses and later, when we rode home in the sunlight, the horses' breath formed ice crystals on their whiskers. The horses were still 'full of beans' and often bucked with sheer joy. We either trotted or galloped most of the way back, and we turned into the gates of 'Willow Hill' just as my parents were running down the drive together to fetch the milk and the morning paper.

Camie was my dearest friend until I met my husband and went to Nigeria, but he was still alive and well when we returned with our children, and I have a lovely photo of my daughter on his back. He lived until he was thirty-two years old.

Shadow, my first Brushtail Possum

One morning, Mr Harvie found a baby Brushtail Possum in the 'Netherby' letterbox. Probably someone had run over the mother on the road and had put the surviving baby in the letterbox, hoping that someone would take care of her. Mr Harvie brought her to me, and I was delighted to rear her. I called her Shadow. When she was old enough to be returned to the wild, we built a wooden box for her to sleep in and fixed it onto a branch of one of the willows just outside the house. I would go out at dusk and she would leap onto my shoulder for a treat of bread, banana or apple. All went well for a year or two, but then she discovered that she could get into our house via the chimney. She traipsed soot everywhere and, on one occasion, she tried to get out through a louvre window and turned into a cubical possum when the panes of glass slid shut under her weight and trapped her against the flyscreen on the outside of the window. After that, my father put a possum-proof screen over the top of the chimney and Shadow moved somewhere else. However, she was found again, living in the fireplace of a house built on the other side of 'Aringa' by some of my cousins. At first, she traipsed soot all over their living room too, but they stopped this by fixing a screen in front of the fireplace. They were very willing to forgo the pleasure of an open fire in the winter for the year-round pleasure of having a tame possum – and her babies – living in the fireplace where they could be observed so easily. However, after a few years, Shadow came back to 'Willow Hill' and took up residence in a box on the shelf of a workshop under the house that was open to the garden by way of the space under the floor of the house. One morning she paid me a wonderful compliment. I had found a way of stroking her that she really seemed to like. I had been stroking her for several minutes when she suddenly grasped my hand in her paws and pulled it under her so that it was near the opening of her pouch. She began to groom my hand by licking and manipulating my fingers with her paws. After licking for about two minutes, she relaxed the sphincter around the rim of the pouch, and then guided my hand in as though it were her joey. Inside the pouch, I could feel a naked, rubbery, real joey about eight centimetres long attached to a nipple. It was an incredible experience! Shadow continued to live in the box under the house and was always friendly. She lived for eleven years.

Collecting things – a life-long passion

During my first year at high school, my mother gave me a book that transformed the next years of my life. It was *Girl of the Limberlost* by Gene Stratton-Porter – the story of a girl who lived in the vast Limberlost swamp-forests of southern Indiana and paid her way through high school by collecting and selling the spectacular caterpillars, cocoons and moths from this habitat. Inspired by this girl, I began to collect moths. It was so exciting going out on summer nights to where moths were attracted, in literally hundreds, to one of the very large, well-lit, plate-glass windows of the house. This was the dining room window and it sloped outwards to reduce reflections. The moths would fly into the bottom of the window, then crawl with fluttering wings to the top before flying out and then down to the bottom again. It was like looking at a waterfall in reverse! Sadly, one does not see many moths at all these days – the 'Silent Spring' prophesied by Rachael Carson has come and my grandchildren know a very different world.

I began to be extremely keen on collecting things and have never lost this passion. Before long, I was also collecting butterflies, beetles and other insects, seashells, fossils, various plants and more or less anything of biological interest. During a school holiday trip to the extinct volcano and caldera at Tower Hill in the Western District of Victoria, I found the skeleton of a dog. Alas, poor Yorick! His bones came home with me and, with the aid of a good book, fencing wire to support the backbone and carpenter's glue, I stuck his skeleton together and mounted it on a block of wood that I kept at the foot of my bed.

I do not know if it were the presence of Yorick in my bedroom or not, but my parents replaced the garage under the house with a carport and then converted part of the garage and my father's workshop adjacent to it into a room of my own that became my museum. In those days, old colonial furniture made of the beautiful Australian red cedar was out of fashion and, when my great aunts died, a glass-fronted bookcase, a chest of drawers and a chiffonier came to me instead of going to the tip. The chest of drawers was filled with sheets of canite onto which my insects were pinned. The glass-fronted bookcase and the chiffonier were filled with jars of specimens in methylated spirits and fossil-bearing rocks collected from many Victorian localities. On the wall were boards onto which my collection of seashells was mounted, and dear old Yorick stood proudly in the middle of the floor.

Rescuing a trapped dog

One day, during my high school years, I rode Camie down the hill through the bush beside Canterbury Road and then a kilometre or so down another road into uninhabited bushland. There, in desperate straits, was a little black spaniel dog with a front leg caught in a well-staked rabbit trap. Without stopping to think, I dismounted from Camie, and approached to comfort the dog and calm it down. I managed to drag the trap's stake out of the ground, but I was not strong enough to open its wicked, serrated jaws. So, I picked up the dog as carefully as I could and, while leading Camie, carried it home. It must have been in agony all the way! As I struggled up our long drive, my grandfather (who lived at 'Willow Hill' in a caravan) saw me coming and told me to stop and stay very still. Then he took off his jumper and threw it over the dog's head and only then did he open the jaws of that horrible trap. Grandpa soundly chastised me for handling an injured animal without taking precautions to prevent getting bitten, but I learned from that dog that desperate animals often seem to know when they need someone to help them and, if one can read their expressions, it is possible to know that they will not bite no matter how bad their pain.

'Glandore' – a lifetime of ecological changes on a Victorian farm

Every year, we had several visits to 'Glandore', a farm on the Goulburn River north-east of Melbourne and on the northern side of the Great Dividing Range. My father grew up on 'Glandore' but then it was acquired by Stu McCracken who had previously lived next door on another property. 'Glandore' was a 160-hectare farm on which sheep were raised for both wool and meat when I first went there but, later, it was stocked only with beef cattle. A creek ran through 'Glandore' and into the Goulburn River that formed one boundary of the farm and, across a stretch of lower-lying ground, there was a string of billabongs or ox-bow lakes that were always referred to as the lagoons. Visits to 'Glandore' invariably entailed rambles all over the farm and especially over the flats with the lagoons. Looking back on all the years that I knew 'Glandore', what strikes me most are the changes that took place in just sixty years. When I first walked around with Stu and my parents, there were Platypuses and Water-rats and many Long-necked Tortoises in the lagoons, together with many various waterbirds and also freshwater mussels, and there were Murray Cod in the Goulburn River that flooded in the winter when the rainfall was heaviest. There were also Tiger Snakes everywhere, including in the outside loo! In fact, the locals used to say, "There's a tiger under every tussock!" There were Eastern Grey Kangaroos and Common Wombats on the nearby Yea Highlands but not on 'Glandore' itself, and there were also no Koalas. Then, in the early 1950s, the Goulburn River was dammed upstream by the Eildon Weir and afterwards the Goulburn was usually very low during the winter, and maximally full – of very cold water from the bottom of Lake Eildon – in the summer. The Murray Cod disappeared and, as time passed, the Tiger Snakes almost disappeared too, and their place was taken by Eastern Brown Snakes and a few Red-bellied Black Snakes. Platypuses became extremely uncommon and, eventually, Water-rats also were very rarely seen, and the mussels disappeared. In contrast, wombats

17

and kangaroos moved down from the Highlands and became increasingly common on the farm, and then Koalas arrived. Probably there had always been a few Koalas in the Highlands, but a sexually transmitted disease – *Chlamydia* – had decimated Koala populations throughout most of Victoria and they did not make an appearance on 'Glandore' until some populations developed resistance to this disease. All of these changes took place without any major changes to Stu's methods of farming. It makes me realize how much things have changed in my lifetime and that the countryside that I knew as a child is a place my grandchildren will never know.

Some changes are natural and do not really matter, but I find it very sad that my grandchildren will not see hundreds and hundreds of moths at their windows at night, and that they should no longer collect all the things I loved collecting when I was young because these things now need all the protection and conservation that we can give them.

School Years

During these first nine years of my life at 'Willow Hill', I attended Heathmont Primary School in the bush, where the headmaster was very enthusiastic about nature studies, and then I went to Ringwood High School for six years. The first three years at high school were easy, but I did not pay much attention to the subject I was least good at – maths. This was very nearly a disaster. I failed maths in year four and this should have meant that I could not do any of the science subjects in years five and six that I needed in order to study zoology at university. Luckily, some very compassionate teachers – including the maths teacher – knew how keen I was to be a zoologist, and they said that I should be given a chance to continue with maths and science in years five and six. These wise teachers realized that abilities mature at different rates and, above all, that having a goal, and a strong incentive to reach that goal, is the best guarantee of success. I am so grateful to them.

Somehow, I managed to pass my year six exams and was accepted by both the University of Melbourne and Monash University to start a science degree. I chose Monash, and another stage of my journey among animals began.

3

DREAMS COME TRUE AT MONASH UNIVERSITY

Monash University was very new – in 1964, the Department of Zoology and Comparative Physiology was only three years old. And this department, unlike the Zoology Department at the University of Melbourne, focused on whole-animal studies such as ecology, physiology and animal behaviour, and the staff included entomologists, limnologists (who studied life in freshwater lakes and streams), two ethologists (who studied animal behaviour), a mammalogist (who worked on Echidnas), a palaeontologist (who studied vertebrate fossils), a geneticist and an invertebrate physiologist. There was no doubt in my mind that Monash was the place for me.

Professor Jock Marshall and his wombat

In 1964 – my first year as an undergraduate – I took courses in physics, chemistry, English literature (because all science students had to do one humanities subject) and biology – and biology was, of course, the subject I loved best. Most of the first-year biology lectures were given by the foundation professor of the 'Zoology Department', Professor Alan John Marshall, known to everyone as Jock Marshall.

Jock Marshall (1911–67) was one of the most outstanding and inspirational people I have ever known. He lost his left arm as a result of a shotgun accident when he was about sixteen but, although he never wore an artificial limb of any kind, he became a commando during the Second

World War, and he led a patrol known as 'Jockforce' deep into enemy territory in New Guinea. And he took mist-nets for catching birds with him. Although most ornithologists and bird-banders have considerable difficulty setting up mist-nets and then disentangling the struggling birds that get netted, Jock could manage single-handed – literally!

He had *such* charisma! He would stride into a lecture theatre, always wearing a dusty, old, black, undergraduate gown – full of gusto and showmanship but *never* arrogance. He would often begin a lecture with a comment on something he had been doing. He had a great sense of mischievous humour. I remember him telling us about a female wombat he had raised and then released into 'Snake Gully', an eight-acre reserve for zoological studies on the Monash campus. Jock attracted many very eminent visitors from overseas and he would prime them with stories about this extremely friendly wombat. Invariably, they all wanted to meet her and they also wanted to see 'Snake Gully', which was the first reserve of its kind in Australia, and so of course they were taken there. They would find the wombat grazing with her back towards them, and she would completely ignore them, even when they attempted to pat her. Eventually, they would feel very misled about this wombat, and they would turn their backs on her. A grave mistake! She would hurl herself at their legs and send them flying as if they were skittles in a bowling alley! Then she would be all over them, as friendly as a puppy!

Jock was an inspirational lecturer who gave us a wonderful introduction to the general biology of animals. And he was a superb judge of people. When choosing people to join his department, he selected only those who he knew would work well together. That was more important to him than their academic prowess. As a result, Jock's department was the happiest workplace I have ever been in or heard of. The tea room was always packed with the lecturers, technical staff and graduate students, and everyone talked to everyone else during this mid-morning break. Ideas were exchanged and debated, and competition and jealousy were non-existent. If someone had a brilliant idea, it was bandied around and taken up by whoever was best suited to do the appropriate research. That tea room was the most important room in the department but almost equally important was 'The Vicarage' – the nearest pub – where staff and students alike would meet and drink beer together after work. Many ideas and research projects were hatched at that famous 'Vicarage'.

In 1965, when my peer group was in second year, Jock became ill with cancer and in 1966 he came into the university only rarely. However, he accepted an invitation to be the guest speaker at a meeting of the Zoology Society although, instead of talking to us, he showed us a film of an expedition he had led from Oxford University to Jan Mayen Island, in the Arctic, in 1947. We saw Jock in that film as he had been in his heyday, and I don't think any of us realized at the time that showing us that film was his swan song. As his health failed, we were grief-stricken and I think many of us would have given our lives to save his.

Second and third years as an undergraduate

In my second year as an undergraduate, I took zoology and physiology as my two major subjects, and biochemistry as a minor. This year, for the first time, we did our practicals in the main zoology building and we were free to move around that building at will. The second-year laboratory for our practical classes became a meeting room in which we could study, talk, play guitars and sing whenever we liked. That was wonderful and so too was the whole zoology building. Jock had designed the second-floor corridor (onto which the labs opened) to have a large open area midway along it. This was lined on three sides by museum cabinets containing displays of all sorts of interesting specimens and models of anatomical structures. Marvellous for revising things we had been taught in lectures and practicals, and for inspiring questions for our lecturers. Also, there was a row of very large glass aquaria or terraria. One contained two living Tuataras – lizard-like reptiles from New Zealand. These are the only surviving members of an order that flourished around 200 million years ago, they have a 'third eye' and they are of great interest to people studying the evolution of reptiles. Another tank, half filled with water, housed an Australian Lungfish, a species belonging to a very ancient subclass of bony fish that originated in the Devonian period about 360 to 420 million years ago. As well as gills, lungfish have a lung (or two lungs in African species) with which they breathe air and this enables them to survive, when pools dry up, by burrowing into the mud. They also have two pairs of lobe-like 'fins' that are similar in structure to the legs of amphibians, reptiles, birds and mammals. Remarkable creatures indeed! However, the tank I spent most time watching contained a small colony of Spinifex Hopping-mice. They reminded me of Winky Jerboa and the animals whose paintings fascinated me most in Ellis Troughton's *Furred Animals of Australia* – the book I had first looked at when I was about eight or nine years old. I just loved these fascinating tiny rodents who hopped around like kangaroos on their long hindlimbs.

The second-year zoology course was mainly an introduction to every class of invertebrates, followed by an introduction to the skeletons and skulls, digestive tracts, respiratory systems, kidneys and renal physiology, and reproductive biology of all of the major groups of vertebrates. Such courses are out-of-fashion today, but I think they provided an essential background of information and that specialising in narrower fields can come later. In my third year, zoology was my only subject, but it included a series of courses on evolution, animal behaviour, limnology, embryology, genetics and cell biology, palaeontology and biostatistics. At the end of this year, I passed the exams for a Bachelor of Science degree and, together with seven of my peer group, I decided to do the extra year required for a B.Sc. with Honours.

Spinifex Hopping-mice – the dainty mice of my childhood dreams

That was a wonderful year! We had to choose four courses from a list covering many topics, but we spent most of our time on a research project. Here again, I was incredibly lucky in that one of our lecturers advertised that he was willing to supervise an Honours student who wanted to work on the behaviour of the Spinifex Hopping-mouse. The thought of having my own hopping-mice thrilled me, and I jumped at the chance to work on them! I decided to produce an ethogram – a written description of all the natural behaviour of these animals. So, I built some observation cages made of wood with sloping Perspex fronts to reduce reflections. Each was 90 × 40 × 45 centimetres in size and was furnished with sand and a few stones and bits of vegetation to resemble the natural habitat of the hopping-mice. Each observation cage had a small nest-box attached to each end, with a hole leading into the cage. Each nest-box had a Perspex front so I could watch the animals in their nests, but the Perspex was usually covered by a wooden shield so it was dark inside the nest-boxes most of the time. Another observation cage had an artificial burrow with two nest-chambers, made out of plaster, which also had a Perspex front that was normally kept covered so the burrow was dark. I watched, photographed and filmed hopping-mice in these observation cages. Sometimes, I had only one animal in the cage but, more often, I staged encounters between several animals.

Spinifex Hopping-mice are found throughout most of the desert regions of central Australia – especially where there are wind-blown sand dunes and clumps of spinifex. The main predators of hopping-mice are owls, and this has influenced many aspects of their evolution – they need to be camouflaged, have eyes good enough to see owls approaching, ears good enough to hear owls approaching, and locomotion that is fast enough and erratic enough to enable them to dodge the owls and avoid capture. Consequently, they are sand-coloured on top (although paler underneath), with very large eyes for a mouse, very long ears for a mouse, very long hindlimbs so they can leap along very rapidly, and a very long tail (with a tuft at its end to increase its air-resistance) that can be used as a rudder and counterweight. Their fast locomotion is similar to that of fast-moving kangaroos.

I observed and described as many aspects of the behaviour of the hopping-mice as possible – locomotion, postures, grooming, feeding, urination and defecation, sleeping postures, glandular marking to spread scent from various glands onto the ground, and all of their social interactions. I also described how they constructed and maintained their burrows, and I found their behaviour in the artificial burrow particularly intriguing because they often plugged the two entrances to each nest-chamber with sand. This made the air inside very warm and muggy, which helped the occupants to save energy and also to save water by reducing evaporation from their respiratory tracts and skins – as Tony Lee and Dick MacMillen proved. After a while, the air inside a plugged nest-chamber had a high concentration of carbon dioxide, but the haemoglobin of rodents that live in burrows is specially adapted so that the oxygen it carries is released into the tissues even when carbon dioxide levels are higher than humans and most other mammals can tolerate.

Spinifex Hopping-mice are very social animals that live in groups comprised of several adult males and females, and their young. They communicate with postures that they can see, with vocal and non-vocal sounds, and with scents from various glands and also urine. Sniffing towards each other, and sniffing while touching each other's noses, mouths, throats, necks, perineal regions and tails, occurs very frequently, especially between group-members. I became particularly interested in all the social interactions between individuals that were caged together, and I ran experiments to observe encounters between two males, or two females, or a male and a female. I left some pairs together long enough to observe courtship, mating and the raising of young. Many of these observations were repeated later, as part of a comparison of social organizations in Australian rodents from very different sorts of habitats (see Chapter 5).

I wrote up my ethogram of this species of hopping-mouse and in 1971 it was published in *Zeitschrift für Tierpsychologie* – a well-known German journal that specialized in animal behaviour. It was published under my maiden name, Meredith Stanley.

At the end of the year, I graduated with a B.Sc. (Honours) degree. This enabled me to start a Ph.D. project at Monash, or an M.Sc. qualifying project at either Melbourne University or the new La Trobe University. It also won me a Commonwealth Secondary Scholarship to any of these universities. For several reasons, I was persuaded to transfer to La Trobe University. I was very sad to leave Monash and all my friends there, but my prospects at La Trobe looked very promising – and so began an interesting interlude in my journey.

4

SAND DUNES, CLAYPANS, KULTARRS AND MORE HOPPING-MICE

I loved the Spinifex Hopping-mice and thoroughly enjoyed writing an ethogram for them. Then I began to wonder if all species of hopping-mice were similar and I also wanted to know if the Marsupial Jerboa, or Kultarr, which looks very similar to hopping-mice and lives in similar habitats, behaved in the same way. The opportunity to compare the behaviour of two species of hopping-mice and the Kultarr came when I went to La Trobe University – the third university to be built in Victoria. There, Dr Patricia (Pat) Woolley was studying the reproductive physiology of carnivorous marsupials belonging to the family Dasyuridae, and these included two species from the edge of the Simpson Desert in south-west Queensland, where there were also hopping-mice and Kultarrs.

My first field-trip to Sandringham Station in south-west Queensland

In June that year, Pat and I took 'the milk-run' flight into the outback country of South Australia and south-west Queensland, and this was my first experience of an Australian desert. The flight was in an old DC-3 aeroplane and it was called the 'milk-run' because it stopped at several cattle stations en route to deliver mail, vegetables and other supplies, and sometimes passengers. In the late afternoon, the plane landed at Birdsville, a small settlement, with a hospital run by the Royal Flying Doctor Service, which is close to the south-west corner of Queensland, where the borders of Queensland, New South Wales and South Australia meet. Pat and I were the only passengers travelling beyond Birdsville, so we stayed overnight at the famous Birdsville Pub together with the pilots, airhostess and groundsman from the plane.

The next morning, we flew into the Channel Country, so-called because it has parallel wind-blown sand dunes running from the north-west to the south-east, with flat claypans, dry riverbeds and gibber plains between them. These dunes channel the floodwaters that come down after heavy rain in the north and on the western flanks of the mountains of the Great Divide in Queensland. The riverbeds fill and overflow and the floodwaters spread out over the flats until vast areas are under shallow water. If the floods are large enough, the floodwaters eventually reach Lake Eyre, a very large inland salt lake in South Australia. Floods were coming down through the Channel Country in June that year, and we started flying over them near Roseberth Station in Queensland. We could see vast flocks of pelicans on the water and also hundreds of ducks, swans and other waterbirds. A wonderful sight!

The runway and homestead of Roseberth Station were on high ground but were completely surrounded by the flood. We landed and dropped off the mail and supplies, and then took off and flew north towards Bedourie, which was our destination. We continued to fly over floodwaters for a while but, as we went further north, it was clear that they were already receding from this area and much of the land was already drying out. The pilot spotted a herd of brumbies – feral horses – and then we were flying over that part of the Channel Country that was to become my study area, and both Pat and I were able to take aerial photographs.

We landed at the small outback settlement of Bedourie and were met by Reg Arthur who was the manager of Sandringham Station, a property of about 5,000 square kilometres that extended into the Simpson Desert. Reg drove us over a track to the homestead – a journey of about eighty kilometres over flat gibber plains and claypans, with some sand dunes running roughly parallel to the track. Gibbers are brown, iron-rich stones that have been rounded and polished by wind-blown sand. When wind has blown all the sand and dust away, the gibbers form a flat plain that appears to be smoothly cobbled. Not much grows on gibber plains. The claypans are also flat – the result of clay and sand being deposited by floodwaters. The claypans support grasses and a few perennial forbs that are usually

26

withered and sparse: the average annual rainfall during the eleven years prior to my trip to 'Sandringham' was only 104 mm and the range was 43 to 201 millimetres, which is not much rain! What really bring the Channel Country to life are the floods that usually occur about once in ten years. When Pat and I arrived on 'Sandringham', the floods had more or less passed but the ground was still damp, there were many pools of water lying about and a green tinge of new vegetation was spreading over the flats.

We came to the homestead and received a very warm welcome from Reg's wife Marj, Colleen, their grown-up daughter, and Barry, their five-year-old son. They all knew Pat from her previous trips to 'Sandringham'. The Arthurs let us use their spare bedroom and we had all our meals with them – usually braised steak, toast and gravy for breakfast, roast beef and vegetables followed by 'brownie' for lunch, and soup, cold cuts of the beef and vegetables for 'tea'. Most enjoyable! I explored the surroundings of the homestead as soon as possible. The Arthurs said I could go anywhere, so I did. Near the homestead, there was a bunkhouse with several rooms where temporary extra hands on the station could be accommodated, and Pat and I used this as a laboratory. Its front windows overlooked the partly roofed yards where stock horses could be kept prior to mustering, or when they were being prepared for the local races (including the famous Birdsville races). There were several horses in the yards when we were there and I longed to ride them, but Reg had a poor opinion of the riding abilities of 'city-slickers' and would not let me. I met some of the more-or-less permanent ringers (stockmen) – some were aboriginal and some were of Caucasian origin – and I came across the herd of goats that supplied the station with milk. Then I was fiercely barked at by Blue Heeler cattle dogs, which have Dingo blood in them – they were always kept tied up unless they were out working cattle. After that, I looked into a large shed with an open door and discovered a Blue Heeler curled up over a litter of puppies in a box. I approached slowly and told her how clever she was to have produced all those puppies and please would she show them to me! She seemed to understand and did not mind at all when I patted her and the pups, and then lifted some of the pups to cuddle them. Dogs may not always be friendly, but they always know when people love them and that can influence their behaviour. Later, the Arthurs warned me (somewhat belatedly) not to go near the bitch in the shed because she was very savage – and especially so now because she had just had puppies!

In the late afternoon when it was cooler, Pat and I went out into the desert in one of the Arthurs' old Land Rovers. Young Barry had the responsibility of making sure there was drinking-water in the car, and checking petrol, oil, tyre pressures and everything else that needed to be checked before it was safe for us to go into the desert, and he was absolutely reliable. It was not his fault that the vehicle had no brakes! We miraculously avoided crashing into closed gates, and eventually set a long line of live-traps across the claypans and gibber plains near the track leading to the homestead from Bedourie, and another at right angles to this line, and then we returned to the homestead at dusk. Early next morning, we checked these 'A-line' traps – always a very exciting task – but this time we had caught nothing at all!

Consequently, we had some time on our hands, so we explored the flank of a dune and found a peaceful pool of water left behind by the recent flood. The still, windless surface of the pool was disturbed only by the slow movement of a solitary shield shrimp that I flicked out of the pool with my hand. We were very surprised then because the water boiled and bubbled as a million tadpoles, other shield shrimps and other crustaceans that had been lying motionless on the bottom suddenly came to life! We collected specimens that we brought back to be identified later by the experts at Monash University.

Pat and I then climbed a sand dune and wondered at the vast number of tracks that indicated the over-night activities of the desert wildlife – carnivorous marsupial Crest-tailed Mulgaras and Kowaris, insect-eating Kultarrs and/or hopping-mice, whose tracks are often similar, and snakes, lizards, centipedes, caterpillars, beetles and even Dingoes.

After lunch, Pat and I drove along a track west of the homestead that led to a distant bore. This took us to what was to become one of my main study areas. It was a claypan bounded on one side by a creek (usually dry) with coolabahs and several other species of trees growing along it. On the other side, there was a fairly stable sand dune with much loose sand but also some mounds of sand that were solidified by the roots of annual legumes and a perennial grass that forms tussocks. We found tracks of hopping-mice on the sand dune and in sandy patches on the claypan, and we also found burrows presumably dug by hopping-mice on the dune and on the claypan. That night, one of the ringers drove us out to this area and Pat and I ran after, and hand-netted, four hopping-mice, which was very encouraging.

The following morning, we again checked Pat's trap-lines and caught nothing, so we moved some of the live-traps to places that looked more promising, including a sandy wash-away that was covered with tracks of Kowaris and other small mammals. And, that afternoon, we set a grid of live-traps, forty paces apart, on what we called B-grid – the dune and claypan west of the homestead where we had caught hopping-mice the previous night.

Our third check of the live-traps along A-line was a breakthrough. We trapped five Kowaris, including two recaptures of individuals who had been caught and marked by Pat six months previously, and one Kultarr. All traps containing animals were brought back to our 'lab' and, there, Pat examined the reproductive condition of her study animals – the Kowaris – while my Kultarr was caged in readiness for bringing it back to La Trobe University. Late in the afternoon, we took the Kowaris back to the trap-site where each had been caught, and then let them go. All of the traps were then re-baited and set again, and then we returned to the homestead for a very welcome bath in an outdoor concrete tank (bogey) into which gushed gloriously warm, slightly saline water from an artesian bore that never stopped flowing! Water from this bore flowed down a narrow channel and then spread out to water an area of lush grasses and sedges where we often saw Brolgas – those graceful, pink and grey, stork-like birds that dance so elegantly when they are courting.

In the days that followed, we checked the traps every morning, re-baited them every evening, studied and cared for our captured study-animals in the 'lab' during the day, and drove out at night to other areas of claypan and gibber plain to see what animals could be spotted with head-torches and the headlights of the vehicle. We could easily have become lost, so one of the ringers always came with us. We saw Kowaris, Kultarrs and many hopping-mice of two species – Spinifex Hopping-mice, which I knew from my Honours study, and Fawn Hopping-mice, which were new to me. Hopping-mice very rarely go into live-traps, so the only way I could collect enough for my study was to run after them with a hand-held net that was like a short-handled butterfly net. I chased one across a bare claypan while Pat followed in the vehicle and recorded our speed. This hopping-mouse bounded along at a steady sixteen kilometres per hour and covered 800 metres before I was able to chase it into a patch of herbaceous vegetation that slowed it down enough for me to catch it. The hopping-mouse was much less puffed than I was, and I was glad that I could catch some without having to run so far! Sometimes I chased and caught Kultarrs too. It was almost impossible to tell them apart when they were racing away at high speed although the hopping-mice hopped bipedally while the Kultarrs always touched down with their forefeet before kicking off with their hindfeet into their next long leap.

Every day brought something new to do. We attempted to dig some hopping-mice out of their burrows on the sand dunes, but this proved impossible. I collected plants from all of the study areas and pressed them so botanists could identify them. I also collected rocks, minerals and fossils from the study areas and copied out the station's rainfall records.

It was especially exciting to be working on Sandringham Station so soon after it had been flooded. Before our eyes, the grasses and herbaceous plants grew to about thirty centimetres high, and many plants began to flower. And suddenly there was a plague of armyworms – the black caterpillars of a moth. While we were still there, the armyworms burrowed into the sand or clay, and emerged two weeks later as moths that gathered in dense clouds around the lights and windows of the homestead. Both species of hopping-mice, as well as the Kultarrs, fed on the armyworms – a protein bonanza too precious to miss. Pat and I also had a marvellous time continuing to investigate the pools that were left over from the floods. When we first arrived, these pools were full of tadpoles (presumably those of the Water-holding Frog), water fleas, fairy shrimps, shield shrimps (which have an almost circular shield over their heads, thoraxes and legs, and a long fork-tipped tail) and two species of clam shrimps – a type of crustacean known from the fossil record as far back as the Devonian period. At first, all of these creatures were fairly small but, by the time we left, two weeks later, many of the tadpoles had already turned into froglets, some of the clam shrimps were more than a centimetre in diameter, and the shield shrimps were, from memory, about seven centimetres long. By this time, many of the pools were little more than puddles of mud and they were packed almost solid with the tadpoles and shrimps. Others had dried out completely and were marked only by mats of dead and desiccated creatures. It was sad in a way, but it was a perfectly normal event and we knew that some of the frogs and froglets had already burrowed deep into the ground where they would survive in a dormant state for even as long as ten years until the next flood. The shrimps, too, were adapted to these desert conditions: they had laid eggs before they died, and these eggs would last until the next flood. Then they would hatch and the cycle would begin all over again.

All too soon it was time to say goodbye to wonderful 'Sandringham' and the dear Arthur family. Reg drove us back to Bedourie where, once again, we picked up the milk-run flight, but this time our destination was Brisbane. From Brisbane, we flew in a much larger plane to Melbourne, where someone met us and drove us back to La Trobe. Our study animals – my hopping-mice and Kultarrs, and some Kowaris for Pat – travelled with us in the cabin, in small holding-cages. By the time we reached home, the hopping-mice had already made neat nests out of the dried grass in their cages, and two of the females had given birth to babies that were being meticulously groomed and suckled.

Comparing the behaviour of hopping-mice and Kultarrs

I had large observation cages at La Trobe, each with about thirty centimetres of sand in the bottom into which the animals could burrow, and pieces of bark under which they could hide. The hopping-mice burrowed but the Kultarrs only hid under the bark. The cages were in a room where the day-night regime was reversed, which meant that I could watch the nocturnal activities of the animals during my daytime. The cages were lit with dim red light that the animals cannot see. It was very satisfactory and my first six months at La Trobe was happy and productive. However, the second half of that year was far from happy. Most of my Kultarrs became sick and died and no vets or medical pathologists could diagnose the problem. I could not visualize anyone getting a Ph.D. by spending all the time trying to keep sick animals alive, and I wanted to change the direction of my research. To cut a long story short, it was decided that I should return to Monash University.

Fortunately, I was allowed to take home one of the surviving Kultarrs – a female who had given birth to minute, almost embryo-like babies that she carried in her pouch like any other marsupial. I put her into one of my old observation cages and recorded her maternal behaviour and also the behaviour of the young from the time they first began to leave the pouch. In 1972, the results of this study were published in the journal *Australian Mammalogy* and, sadly, this was the only paper I was able to write from the work I had done at La Trobe University.

5

SOCIAL BEHAVIOUR OF NATIVE RODENTS FROM DIVERSE HABITATS, LITTLE BANDI AND POTOROOS

I was very relieved to be back at Monash University and I was permitted to start immediately on a Ph.D. project, under the supervision of Dr John Nelson. My idea for a project was inspired by the work of Tony Lee, Dick MacMillen and Russell Baudinette (one of my peer group), who had been studying Spinifex Hopping-mice, Fawn Hopping-mice and Sandy Inland Mice, all of which live in central Australian deserts where temperatures and humidities above ground are usually intolerable during the day, food is often and very unpredictably scarce, water for drinking is very seldom available, and succulent vegetation is also scarce except after rain or the floods that sometimes occur only after intervals as long as ten years. It was known that these desert mice avoid the hostile surface conditions by digging deep burrows in which they spend the daytime, and Tony and Dick had discovered that they can survive without drinking water or eating succulent plants. They can do this because they have extremely efficient kidneys that can excrete highly concentrated urine and they are able to obtain metabolic water from the carbohydrate-rich seeds they eat by converting the carbohydrate to water and carbon dioxide. Nevertheless, these desert mice must conserve water in every possible way, and they do this in several ways including by producing very dry faeces and by not sweating. Another way is by huddling together in compact heaps so that each animal has less of its surface area exposed and consequently loses less water from its skin as a result of evaporation. Tony Lee and his team found that huddling has an additional advantage – the metabolic rate of huddling animals is lower than that of the same animals measured individually. Therefore, huddling animals expend less energy and require less food, and this is especially advantageous during long periods of drought when food is scarce. This research revealed that, because of these benefits of huddling, it was advantageous for small rodents living in deserts to live in groups. I already knew from my Honours project (Chapter 3) that captive Spinifex Hopping-mice exemplified a lot of behaviour that resulted in the formation and maintenance of groups, and I knew from the work of Tony Lee and his colleagues that Fawn Hopping-mice and Sandy Inland Mice also formed groups in captivity. This raised questions that interested me enormously. Did *all* Australian desert rodents live in groups? And what about those that lived in semi-deserts, heathlands and other non-desert habitats? Suddenly, I had my inspiration for a Ph.D. project.

I was very lucky. I already had Spinifex Hopping-mice and Fawn Hopping-mice to work on, and also there were still some Sandy Inland Mice at Monash and I was able to obtain some Plains Mice from colleagues in South Australia. All of these species live in central Australian desert habitats. Furthermore, I had met Clive Crouch during my Honours year and I knew that he would help me to obtain Silky Mice and perhaps also some of the rare Mitchell's Hopping-mice from semi-desert habitats in the Victorian Mallee. And Joan Dixon, the curator of mammals at the National Museum in Victoria, knew where I could obtain Heath Rats, a species that lives in heathlands in Victoria, and Peter Aitken, the curator of mammals at the South Australian Museum, was willing to take me to an oasis habitat in central Australia, where he had discovered an amazing assemblage of small mammals that included Desert Mice, about which almost nothing was known at that time. Studying differences in the social behaviour and social organizations of these rodents would have the great advantage that they are all members of the Conilurinae, a subfamily thought to be descended from a single invading stock that reached Australia in the Miocene (possibly as long ago as fifteen million years) or perhaps only a little later in the Pliocene. This meant that similarities in their behaviour were likely to be a consequence of their descent from a common ancestor, while differences were most likely to reflect adaptations to different habitats. Similarly, differences in their reproductive strategies (litter-sizes and the timing of reproduction in relation to changes in the habitats) were also likely to reflect adaptations to different habitats.

As my study evolved, I became particularly interested in the extent to which aridity and hostile conditions above ground influenced the evolution of social behaviour in the conilurine rodents. I also became equally interested in the extent to which the stability or otherwise of the carrying capacity of each habitat influenced the evolution of social behaviour – the carrying capacity of a habitat for a particular species is the maximum number of individuals that the habitat can support at any particular time. It was with these interests in mind that I established several study areas and, as well as observing the behaviour of my study animals in the field, I measured temperatures and relative humidities in the microhabitats and domiciles exploited by the animals, and also in those places that were avoided. Also, I obtained rainfall records and collected plants so that I could assess the availability of food. I had four main study areas – Sandringham Station in the Channel Country for the desert species, the Big Desert of Victoria for the semi-desert species, Pomonal near the Grampian Mountains in Victoria for the heathland species, and Johnson's Bore near the north-west shore of Lake Eyre in South Australia for the oasis species.

Sandringham Station for the second time

I needed to go back to Sandringham Station to make further observations on the habitats of Spinifex Hopping-mice and Fawn Hopping-mice, and Geoff Hutson, one of my fellow Ph.D. students, wanted to go there to work on the behaviour of Kowaris and Fat-tailed Dunnarts, which are marsupials. Three other Ph.D. students wanted to come too – partly to assist us with our work and the driving, and partly to collect samples of water and plankton from springs and bores in the outback, and to make tape-recordings of the songs of Willie Wagtails in many remote localities. This, my second trip to 'Sandringham', took place in September 1969 and it followed my field-trip to Johnson's Bore, with Peter Aitken and Tony Robinson from the South Australian Museum, which is described below.

Having completed our work at Johnson's Bore, we drove to Marree and parted company with Peter Aitken and Tony Robinson. The rest of us, in a Toyota LandCruiser, set off up the notorious Birdsville Track that runs from Marree in South Australia to Birdsville in the north-east corner

of N.S.W. We camped, en route, in the deserted Lake Harry Homestead –
one of the many homesteads of early settlers who were forced to leave and
abandon their cattle when drought conditions prevailed, as they do most
of the time in this part of Australia. Along the Birdsville Track, the biggest
hazard we had to contend with was bulldust – deep drifts of very fine white
dust that were quite difficult even for the LandCruiser to get through. It
was a comfort to know that people at Marree knew we were attempting to
reach Birdsville, and we had also contacted people on all the cattle stations
that we passed through, so our position was always roughly known and
someone always knew when we were expected at our next destination. We
stopped quite often along the track to collect plankton and water-samples
from artesian bores. At one locality, the water in the pool at the bore-head
was 65.5° centigrade, and at another bore the water was 73.5° centigrade,
but both pools supported clouds of thriving plankton.

About 85 miles (1 mile = *c.* 1.6 km) from Marree, we crossed the dry
bed of Cooper Creek, which was five miles wide at this locality. A little
further north, we entered the Channel Country with the long, parallel sand
dunes, claypans and gibber plains that I remembered so well from my first
trip to Sandringham Station. We camped at Mirra Mitta bore and set live-
traps but did not catch anything and, the next day, we crossed the vast sandy
flats of Goyder's Lagoon and came to Birdsville, 308 miles north of Marree.
We pushed straight on through Birdsville and camped fifty miles further
north at Carcoory in the ruins of yet another abandoned cattle station. The
following day, we came to Glengyle Station and met the Crombies who
had lived there for twenty-eight years. They were keen naturalists who told
us that, on Glengyle Station, Fat-tailed Dunnarts were abundant in vast
tracts of lignum that grew on the alluvial floodplains of the Eyre Creek,
and also that seven Stripe-faced Dunnarts had been captured there recently.
The Crombies also reported that Fawn Hopping-mice were abundant in
areas that were more typical of the Channel Country, that Long-haired
Rats occurred on the flats, and that one rare Bilby had been captured
somewhere on the station.

We drove from Glengyle Station to Bedourie and from there to
Sandringham Station, and we were offered the use of a stockmen's
bunkhouse that was, fortunately, unoccupied at the time. The current
manager was Norman 'Peewee' Clark, a very friendly and helpful man.
That evening, we set traps along the Sandringham-Bedourie track and, after

dark, we drove the LandCruiser over claypans and gibber plains north-west of the homestead, and hand-netted several Fawn Hopping-mice, Kultarrs and Fat-tailed Dunnarts that were brought back to our quarters for examination and caging.

We then spent ten days live-trapping, excavating burrows, and recording temperatures and relative humidities at hourly intervals in burrows and just above ground, and also in deep crevices that had opened up as the claypans and gibber plains dried out after the flooding eighteen months earlier. Some of our live-trapped Kultarrs disappeared into these deep cracks when they were released, and presumably that was where they lived during the day. In captivity, they never dug burrows. I copied the station's rainfall records: potential evaporation exceeds precipitation here, and droughts lasting up to ten years are not uncommon. From my records, and also from those of other studies, I confirmed that there were marked diurnal and seasonal fluctuations in shade temperature and relative humidity above ground, and that surface temperatures during the day exceeded the upper limits of tolerance of small mammals throughout the year. Furthermore, low humidities, intense solar radiation, winds and sandstorms were periodically additional hazards above ground. In contrast, soil temperatures a metre below the surface did not fluctuate during the day, and the seasonal range was only about twelve degrees – from about 22° to 34° centigrade – so it was not surprising that the small rodents from Sandringham Station dug burrows into the cool sand and were only active above ground at night.

I spent some nights alone out on the sandy claypan adjacent to one of the sand dunes north-west of the homestead on what had been my B-grid in 1968 and, with the aid of a spotlight, was able to observe hopping-mice and Kultarrs foraging and feeding there. Sometimes, groups of five or six hopping-mice would approach within less than a metre if I made squeaking noises similar to those of their young. In 1968, I had observed that Spinifex Hopping-mice fed and moved around in groups: distances between group-members varied from about one to 100 metres, and different groups were about 300 to 700 metres apart. It proved very difficult to excavate burrows of Spinifex Hopping-mice on Sandringham Station, but burrows at three other localities in central Australia that were excavated by other people, contained five to eight individuals. The occupants of three burrows were identified as (1) a male, two females and two litters (total ten), (2) four adult females and three adult males, and (3) four adult females and two immature males.

The Big Desert in Victoria

My observations of the social behaviour of the semi-desert conilurine rodents focused on Silky Mice and Mitchell's Hopping-mice and I had three trips into the Big Desert in western Victoria, where both species occurred. The Big Desert, and also the Little Desert further south, is warm-temperate, semi-desert Mallee and heathland that lies between the 250- and 500-millimetre isohyets. Mallee gums are species of *Eucalyptus* that have very extensive root systems and multiple thin trunks. They are usually well spaced and have a sparse understory of low shrubs, heaths and grasses that are drought-resistant and lie dormant when the rainfall is low. Although precipitation exceeds evaporation in these habitats, they resemble the deserts further north in that the rainfall is unreliable and droughts are frequent. I followed temperatures and humidities and found that, although the surface conditions were less severe than in the deserts, it was nevertheless necessary for both Silky Mice and Mitchell's Hopping-mice to spend the day in deep burrows that they dug themselves, and they were strictly nocturnal. I was very lucky to have met Clive Crouch, a brilliant naturalist who had been doing a lot of small mammal trapping, in both the Big Desert and Little Desert areas, using very effective, home-made wooden live-traps. At a locality called Pigick, Clive trapped, marked and released the mice he caught, and he routinely followed Silky Mice back to their burrows. He discovered that these burrows usually contained one adult male and one or two adult females, and often the females were either pregnant or suckling young. He discovered that females captured on consecutive nights always returned to the same burrow whether or not they had young in that burrow. In contrast, on consecutive nights, many of the adult males returned to different burrows that had different female occupants. The home-ranges of the males were variable. Usually, the males ranged over restricted areas of the colony and had access to only three or four burrows. However, for periods of one to two months, some of the males ranged over most of the colony and had access to about twenty burrows. Clive also discovered that there were periods when many new burrows were established. During these periods, subadults that had previously lived in their mothers' burrows moved into new burrows. The subadult females lived singly or in twos in burrows that were visited by adult males, one at a time, but the subadult males were found singly in new burrows until they reached reproductive maturity several months later. Adult females usually remained in their old burrows, and they were often pregnant or suckling new litters when the young of older litters left. Those adult females continued to be visited by adult males.

Clive and I excavated and mapped several burrows and found that they were up to 130 centimetres deep and, like those of the desert species, they had large nest-chambers linked by horizontal tunnels, with vertical pop-holes out of which the mice could emerge above ground.

Pomonal

My study area for Heath Rats was an area of flat, sandy habitat at the foot of the Grampian Mountains in western Victoria, and I spent four or five days there every month for a whole year. This habitat was cool temperate, dry sclerophyll forest, with well-spaced eucalypts and an undergrowth of dense to medium-dense heathland flora dominated by the shrubby Heath Tea-tree (*Leptospermum myrsinoides*) and including grass-trees (*Xanthorea* sp.), bracken and other heathland plants. The mean annual rainfall was 720 millimetres and, although it varied from year to year, there was always enough rain to promote the growth, flowering and fruiting of the tea-trees and other plants eaten by Heath Rats, and water was always available from plants, rain or dew. Therefore, although the carrying capacity of this habitat for Heath Rats has seasonal fluctuations – being highest in spring and summer – the fluctuations are comparatively predictable and are much less extensive than those in the desert and semi-desert habitats. Another big difference between the heathland habitat and the desert and semi-desert habitats is that temperatures and humidities above ground do not fall outside the ranges tolerated by small mammals and therefore the Heath Rats are able to spend the day in simple surface nests that they construct under dense vegetation. Heath Rats breed during the spring, when food is most abundant. They invariably have three young per litter and there is probably no post-partum oestrus or early maturity, so the reproductive rate of this species is much lower than that of the other species I studied.

A narrow, unsealed and very seldom-used road ran right through my study area and, in one place there was a wider area where cars – if any – could park. This was a great place for me to pitch my little green tent and put up a folding table on which I could work, and usually my parents came with me on these field-trips and they lived in a caravan that could be parked near my camp. Having a road and campsite in the middle of a study area was unconventional, but I felt confident that it would not affect the results of my study and, indeed, being so close to the habitat of my animals had many unexpected advantages.

I set up a 160 × 225 metre grid of ninety permanent trap-sites. The sites were set twenty metres apart on ten parallel lines twenty-five metres apart. I set live-traps at each site and, by capturing, marking and releasing the animals, I hoped to obtain information about their movements and social interactions, and their reproduction. This procedure only works if the traps provide shelter and, above all, enticing food, so that the animals are willing to go into the traps on successive trap-nights. I baited the traps with a delicious mixture of peanut butter, honey and rolled oats, and most of the small mammals on the grid found this irresistible. During the year of my study, I caught twenty-four Heath Rats and obtained cumulative records for seventeen individuals. Of these, fifteen ranged over comparatively small areas, while two ranged over most of the grid. Some animals were caught at the same sites, or adjacent sites, on consecutive nights, and it was clear that they had overlapping home-ranges and that they had formed associations of some sort. Occasionally, two individuals were even caught together in one trap. The associations were almost always formed between an adult male and an adult female and it became apparent that Heath Rats live in pairs throughout the year. I put small radio-transmitters on two individuals that I thought were a pair from the trapping results, and I tracked them to a nest where they were huddling together.

Little Bandi and other bandicoots at Pomonal

As well as the Heath Rats, I trapped Swamp Rats, Yellow-footed Antechinuses, Southern Brown Bandicoots and Long-nosed Potoroos. On the grid, there were also Koalas, Red-necked Wallabies, an Echidna and occasionally (unfortunately) a few feral cats and foxes.

Bandicoots were trapped quite frequently because they were moderately trap-prone, but one individual became absolutely trap-addicted! I first caught this one – Little Bandi – when she was about half-grown in March and, because she was quiet and easy to handle, I took her away to show her to some of the local residents. She can't have found this too stressful because I retrapped her the next night. In April, I retrapped her again and that's when she started to become addicted to the traps and their delicious contents. I would find her in a trap, shake her out into a cloth bag, weigh her and examine her on the spot, and then let her go – whereupon she would run to the nearest empty trap and bound inside to eat that bait. I caught her nine times one morning while checking the ninety traps on the grid!

Her addiction did her no harm until early May, by which time the nights began to be quite cool. Having been in one trap all night long, and undoubtedly also for most of the previous day, she was invariably cold and probably miserable by the time her trap was opened and she was released. I often felt cold too and wore a thick woollen jumper with long sleeves. One morning, just as I let Little Bandi go, I pulled the sleeve open to form an attractive, warm 'burrow' in front of her nose. She was tempted. She pushed herself into the sleeve and wriggled all the way up to my warm shoulder, where she settled down and stayed until I had finished checking all the other traps. When I returned to my camp, I boiled a billy of water, filled a hot water bottle and put it – and the bandicoot – into my sleeping bag! My tent did not have a sealed-in floor, so I assumed she would easily find her way out when she wanted to. However, to my amazement, she was still curled up asleep when it was time for me to turn in! So, I got into the bag with the bandicoot and she stayed with me for about an hour before hunger drove her outside. A minute later, I heard the door of the nearest trap slam shut! I did not get up to release her – I was not prepared to spend all night following her from trap to trap! But, in the morning, I let her out and again she climbed up my sleeve and spent another day in the warmth of my sleeping bag. This went on for several months but then she was no longer captured.

The bandicoots on my study area did some surprising things. On several nights, at least one of them tried to burrow into my tent. And one cunning individual discovered that it could prevent trap doors closing by digging sand over them. By this means, the bait from a large number of traps could be stolen! Tracks and droppings around the raided traps were clearly those of a bandicoot and the unexpected failure of one adult male to be recaptured after being somewhat trap-prone made me suspect that he was the culprit! But perhaps Little Bandi learned that trick too.

The potoroos at Pomonal

Little Bandi and most of the other small mammals on the study area were trap-prone if not trap-addicted, but the potoroos (with only two exceptions) were the exact opposite and I began to dread catching them. Potoroos are one of the smallest members of the kangaroo family. They stand about thirty-five centimetres high and weigh 660 to 1640 grams. Most of my traps were too small for a potoroo to get into, but sometimes I had to use some larger, wire-mesh traps in order to have a trap at each of the ninety trap-sites, and I occasionally caught potoroos in these. It was not nice! The potoroos almost invariably panicked when they were caught and then bashed themselves against the sides and tops of the traps. They panicked again when I approached the traps, and they ended up with bleeding noses and dishevelled fur.

During one early field-trip, I was visited by a keen and very experienced wildlife photographer who had been told that I sometimes trapped potoroos. He was delighted when we found that two had been caught that morning. His technique for photographing small wild animals was to set up an oval enclosure of corrugated-iron about 1 × 1.5 metres in area, and about 1.3 metres high. He would place the enclosure in suitable habitat and then get inside the enclosure with the animal to be photographed. With the potoroos, this was a dismal failure. As soon as the first animal was set free in the enclosure, it leapt onto Ian's knee and from there onto his shoulder and out over the wall. The second did likewise except that, en route, it managed to bite Ian's finger, almost to the bone. He was not very impressed!

I invited Ian to come back to Pomonal several months later. In the meantime, two male potoroos had decided that being trapped was not such a bad thing after all. They discovered that they were always handled very gently and that their irritating ticks were removed, and they were given many titbits of bait while I was examining them. Unlike all the others, these two males became trap-prone and I could almost guarantee catching them. So, I telephoned Ian. It was not easy to persuade him to come this time, but he *did* come and was somewhat disconcerted to find that I had not yet even checked all my traps. "Don't worry," I said. "There will be a potoroo just over there." Sure enough, when the trap was sighted, there was a little 'pot' sitting up with pricked ears and calmly waiting for us to approach. "Now, just watch this!" I said. I spread a hessian sack on the ground near the door of the trap and, when I opened the trap, the little 'pot' hopped out and sat on the sack. I wrapped the sack around him and then carried him back to my camp and put him on my table. I gave him a piece of delicious bait and then rolled him over so I could check his reproductive condition and remove any ticks. I told Ian that there was no need for him to set up his enclosure. "I can put him down anywhere you like," I said, "and he won't hop away". This was too much for Ian to believe, so he set up his enclosure and both of us climbed inside with the potoroo. "Where exactly do you want him," I said, "and what direction do you want him to face?" Following Ian's instructions, I gently put the little fellow down. I asked Ian if he was happy with the position of the little animal's tail and a rather bemused Ian said, "Well, it might be better if it were moved around this way a little." I moved the tail, and the potoroo did not move. "That's no good," said Ian. "It looks half dead!" "Don't worry," I said. "You get the camera focused and tell me when you are ready." When Ian indicated that he was ready, I clicked my tongue and the darling little animal sat up, pricked his ears and looked at us. Ian got some marvellous photographs!

A unique record: the retrieval of a joey by a potoroo

As I mentioned earlier, almost every other potoroo that I trapped was so terrified by the experience that it never entered a trap again. So, you can imagine how distressed I was one morning in June when I caught two females, each of which had a joey. In dangerous situations, the natural reaction of females in the kangaroo family is to throw their joeys out of the pouch and save themselves. This makes good sense because they will already have another embryo *in utero* waiting to start developing and it is better to save this one and themselves than to risk the death of all three. I tried to put each joey back into its mother's pouch, but this proved impossible, so I decided to put the joeys down on the ground and then release the mothers without distressing them further by weighing and examining them. I was not surprised when the two mothers bounded away, leaving their joeys behind. I went away for about an hour, and then went back to see if the joeys were still there. Unfortunately, they were and, by this time, they felt quite cold. Obviously, I could not leave them any longer, so I took them back to my camp and warmed them up on a hot water bottle. Then I drove to the nearest small town and bought eyedroppers, dolls' feeding bottles and some fresh milk, honey and glucose. Back at the camp, I managed to feed both joeys with the eyedroppers and they began to revive. The oldest was a female and her eyes were open and she had fur that was long enough around her head to show the adult coloration. The youngest was a male and, although his eyes were also open, his fur was only just through and it looked as though he were made of black velvet. The little female was able to hop quite well, but the male was very wobbly on his feet. However, he made some progress and then both joeys tried to follow me around the caravan during their short exercise periods. They also followed each other, but they spent most of the time trying to find a pouch exactly where a pouch would have been had each joey been an adult female. They also tried to suck – and then bite – anything that projected outwards such as each other's ears and corners of the cloths I tucked around them. The poor little male suffered greatly because he had a very prominent projection in exactly the right place!

I kept telling myself that the joeys could not possibly survive for long, but they survived the first day (in the caravan) and the first night with me in my tent and, after that, I was more hopeful.

I kept the joeys in a cardboard box with a hot water bottle and warm cloths that they could snuggle into. At night, in my tent, the loud, sibilant hissing calls they made when they were hungry woke me up and it only took a few minutes to warm their milk and then feed them. But, on the second night, a very remarkable thing happened. I started to feed the little female first, and this caused the male to call with even more vigour – and then I heard the thud, thud, thud of an adult potoroo approaching the tent. I recognized the sound because I had released adults and listened to them hopping away many times. The potoroo outside came right up to the tent and then started frantically hopping around it, jumping onto the walls and sliding down, and scratching where the walls were pinned down. The joey responded by making frantic efforts to get out of the box, and it continued to call and call. Then the adult drummed on the ground. I decided that I had to let the joey go, so I turned off the torch that was lighting the tent, lifted one of the walls off the ground, and placed my little friend outside. Then I listened. I heard the adult hop up to the joey, and then there were the sounds of both of them moving slowly away. After moving perhaps two metres or thereabouts, they stopped. There was a moment of silence, and then the adult bounded away at high speed. A minute later, I went outside and shone the torch around, but there was no trace of the joey!

I have no doubt that the joey had been retrieved by his mother but unfortunately, I was not able to prove it by recapturing it with its mother and I was told that I would not be able to publish the story in a scientific journal because my conclusions were based only on what I had heard. This was very disappointing because, as far as I know, this would have been the first and only record of a member of the kangaroo family retrieving its abandoned joey.

The rest of this story is sad. On the third morning, I found that the little female had died. She had appeared to be thriving only an hour earlier when I had last fed her, so I wondered if she had tucked her head into a thick fold of cloth and suffocated.

Johnson's Bore – an oasis in the desert

The habitat that provided the greatest contrast with the desert, semi-desert and heathland habitats was the oasis-like habitat of the Desert Mouse on Anna Creek Station near the north-west shore of Lake Eyre. In 1917, a bore had been drilled there, deep enough to reach the water of the Great Artesian Basin. Possibly the site of the bore had been chosen because there was some seepage of water to the surface, but it is unlikely that there had been an old natural spring at the site because there was no mound of minerals even though the bore water was quite saline. The flow of water from the bore was very substantial and so the South Australian Mines Department tried to control it with a concrete casing and a manually operated tap. This failed. The water pressure was so great that the casing was blown out of the ground and thereafter the flow of water from the bore was about 3.3 million litres per day.

The water flowed down a number of shallow channels that fanned out to create a narrow triangular area of moist sand, about a kilometre long. The moist sand supported dense tussocks of Spiny Flatsedge (*Cyperus gymnocaulis*) that were sometimes entangled with bushes of Shrubby

Samphire (*Tecticornia halocnemoides*), and some of the clearings supported low-growing Red Sand Spurry (*Spergularia rubra*), a species of saltbush, White Paper Daisies and a few grasses. This oasis of sedge was in marked contrast with the flat, arid, saltbush country surrounding the bore.

I went to Johnson's No. 3 Bore in August 1969 with four other postgraduate students from Monash (in a Toyota LandCruiser with trailer attached) and Peter Aitken and Tony Robinson from the South Australian Museum (in a Toyota truck). It was my first venture into central Australia by road, and it was very exciting. We spent the first night camped near the northern limit of the semi-desert country in South Australia, and the next day we drove north across saltbush flats and gibber plains to Andamooka Station. The road was very rough and we drove several kilometres apart to avoid each other's dust.

The next day, we left the road and drove cross-country. Peter knew the area well enough to navigate from one familiar hill to another. We came to the dog-fence and followed it to the Bull's Head Gate, which we could open and drive through. The dog-fence was constructed to keep Dingoes from the cattle country in the north out of the sheep country in the south and it was, I think, the longest fence in the world. It was constantly patrolled and kept in good condition. North of the dog-fence, we passed Mattaweara Lagoon, where we stopped to enable Keith Walker (who was studying life in inland waters) to take plankton samples, and we stopped again to have lunch about four miles north of Pimba Dam. About ten thousand years ago, aboriginal people had made stone tools here, and we collected several macrolithic percussion stones and knapped implements that were lying exposed on the surface.

After lunch, we drove cross-country to Stuart Creek Station, and then joined the Marree-Alice Springs Road that follows the old Ghan railway. The country here was flat and barren and appeared to have been devastated by overstocking. On this leg of the journey, we passed several mound springs where water bubbled up at the top of small hills of accumulated minerals that were deposited when the mineral-rich spring-water evaporated. At the top of a mound spring called Wobna there was a pool of hot water and, every five minutes, mud erupted to the surface and protruded like the exposed bare brown bottom of a diver. Keith recorded that the water temperature was 30° centigrade and that specially adapted fish and amphipod crustaceans were abundant in the pool.

About 100 metres south of Wobna, wind had blown away some of the mineral deposit to reveal yellow sands in which were embedded more aboriginal artefacts. These were microlithic pirri points, trapezes and other knapped implements that were about 25,000 years old. That night we camped near another deserted homestead near Coward Springs, and the next day we pushed on past more mound springs to William Creek. There was a pub there, but I cannot remember any other buildings. We stopped to have drinks at the pub, and there we met a gang of men who were working on the railway nearby. One was an Englishman who was known as 'Mick-the-Duck'! We told the gang that we were going on to Anna Creek Station and they told us that water would be a problem where we were going. "The trouble is – the pipes get blocked by centipedes." We must have looked very startled, so they hastened to add, "Of course it's all right if there's a house on fire. You can squirt them out and use them for ladders!"

From William Creek, we drove north over extremely arid and desolate country that supported only a few stunted saltbushes, and eventually we came to some cattle yards near the head of Johnson's No. 3 bore. We pitched tents and set up our camp under a solitary clump of trees about 1.5 kilometres from the bore-head. The next day, we surveyed the long triangular area of moist sands and then set a combination of live-traps and snap-traps, mainly in or near the sedge. We chose places where there were signs of rodent activity such as runways, droppings, chewed sedge, scratchings or burrows. We excavated one burrow and found it occupied by a Long-haired Rat, a species that forms plagues that spread out across the desert in the wake of floods. When the land dries out, most of these rats perish, but enough take shelter in refugia, such as the oasis around Johnson's Bore, to keep the species widely distributed. This locality was the most south-westerly record of the Long-haired Rat, which has its stronghold in the Barclay Tablelands in the far north.

At dusk that first evening, I set up an observation station in a small clearing surrounded by sedge and prepared to spend the night there by myself. A portagas lamp kept the area well lit until I turned it out at half past ten. During that time, I kept records of any animals that I saw. I noted a few Desert Mice running across the clearing and then disappearing into the sedge. They were always by themselves. I also observed some Barn Owls sweeping over the sedge on wings that most rodents cannot hear, and I asked my colleagues to try to find their roost-sites so that we could collect and examine their pellets of regurgitated fur and bones to determine what they were eating. They only found one pellet, and it only contained a few skulls of Long-haired Rats. My most important task during the night was to record the temperature and relative humidity, at hourly intervals, in some of the runways made by Desert Mice through dense tussocks of sedge. For comparison, at the same time, I recorded the temperature and relative humidity just above the ground in the open clearing. I set an alarm clock to wake me every hour. It was strange being there all alone except for the little animals and the owls. At first, I could see the lights of the others' camp more than two kilometres away but, after they turned in for the night, I could have been a hundred kilometres from the nearest human habitation.

I had a lot to do at Johnson's Bore. The traps were checked at dawn and all of the Desert Mice were taken back to camp. All the standard measurements were made, and then the living animals were caged. That night, after the mice had been without food for twelve hours, we gave them a choice of vegetation from the oasis habitat and recorded what they ate. We discovered that the sedge provided all the needs of these mice – they ate the stems, flowers and seeds. The sedge also insulated the mice from the extremes of temperature and humidity in the open. The mice bit through the stems and removed sections of them to create very narrow runways (3–4 centimetres in diameter) through the sedge. These were too narrow to allow individuals to pass each other but they led out to comparatively wide runways that ran around the bases of the tussocks under overhanging stems. At intervals along the narrow runways, the mice also cut out chambers (8–15 centimetres in diameter) that had a variety of functions. The largest were lined with fibre from chewed-up stems to make very neat spherical nests, about fifteen centimetres in diameter, in the middle of the dense tussocks. Other smaller chambers only had fibre covering the floor, and others contained caches of food such as rhizomes, flowers and seeds of the sedge. I found it very interesting that all of the nests that we examined appeared to be too small to house more than one adult mouse and, when we set several traps around tussocks containing a nest and runways, we only ever caught one animal. Throughout the day, I continued to measure the temperatures and relative humidities in the runways and out in the open, and I collected specimens of all the plants I could find, for later identification.

We spent six nights at Johnson's Bore. I continued to measure temperatures and relative humidities and every spare moment was spent catching animals, housing them, feeding them and observing them. At the end of this time, I was completely exhausted after nights of little or no sleep, and I remember very little of our return journey to William Creek and then Marree, where the Monash team said goodbye to Peter and Tony from the South Australian Museum. They returned to the museum while we set off up the Birdsville Track to Birdsville and then Sandringham Station – but 'Sandringham' was my main study area for the hopping-mice and my experiences there have already been described.

Investigating Social Organizations

Social organizations are formed when individuals in a population repel or attract each other. I classified elements of repelling behaviour (agonistic behaviour) into four broad categories – attacking, pursuing, fighting-in-contact, and nest-defence. Similarly, attracting behaviour (amicable behaviour) comprised contacts for investigation (such as nose-to-nose sniffing), huddling together, grooming each other, and mounting. To compare the social organizations of the different species, I staged encounters between two individuals in observation cages that were furnished to resemble the natural habitat of each species and, attached to each end of a cage, there was a nest-box containing nest material from the home-cage of each animal. Consequently, each animal had a nest-box that represented its own territory whereas the open-cage was neutral ground, unfamiliar to both. The room in which the animals were kept and observed had artificial lighting with the 'night' and 'day' regimes reversed so I could watch the night-time activities of the animals (in dim red light) during my daytime. I recorded the number of agonistic and amicable elements of social behaviour that occurred during seven thirty-minute observation periods per night, for at least ten nights. I also recorded avoidance, mutual avoidance, the occurrence of copulation and the total amount of time that the animals spent together, either in the open-cage or in one nest-box. The participants in the encounters were adult – either two males, two females, or a male and a female. Single-sex encounters, and those between non-breeding males and females, were run for ten days but encounters between males and females in breeding condition were run for as long as it took for

at least one litter to be raised. These encounters resembled soap operas, and
I never became bored by them.

Hopping-mouse soapies

Spinifex Hopping-mice were delightful to watch. They are such graceful
animals with their beautiful large, round eyes, very long ears, very long,
slender hindlimbs and long tail with a pencil of long hairs at its end – and
their social interactions were almost always equally delightful to follow
from day to day.

Encounters between two adult male Spinifex Hopping-mice usually
began with a brief period of agonistic behaviour. During this period, one
male usually asserted dominance by attacking and chasing the other male
but fighting-in-contact was rare so there were seldom any injuries. Very
soon, the animals began to approach and sniff each other and, soon after
that, they began to huddle together, mount each other and groom each
other. This behaviour bonds hopping-mice together and, by the end of
the first night, all aggressive behaviour had ceased and the males nested
amicably together in one of the nest-boxes. For the next five or six nights,
acts of amicable behaviour were particularly frequent. After that, they were
less frequent but there were nevertheless between thirty and forty acts
every half-hour and, from the first night onwards, the males utilized the
open-cage area at the same time and continued to nest together.

Encounters between two females began with much more aggressive
behaviour than was ever shown by the males, and fighting-in-contact
resulted in bite wounds but never fatal injuries. However, aggressive
behaviour usually ceased before the end of the first night, and the females
usually nested together. During the following nights, there was much less
amicable behaviour than between males, and the females spent less time
together because they tended to utilize the open-cage, or rest in their nest-
boxes, at different times.

Encounters between a male and a female also began with aggressive
behaviour but all aggression ended during the second night and, from then
on, there was a steady increase in amicable behaviour – usually contacts for
investigation and huddling – and the pair began to rest together in the same
box, feed together, burrow into the sand together and care for the shared
nest together. When the female came on heat, contacts for investigation,

social grooming and mounting increased in frequency while huddling decreased. When she became receptive, the male followed her most of the time, sniffed her genital region, and often mounted her and groomed her, especially while mounting her. It appeared to be a very loving courtship between two animals that were already strongly bonded together. The period of intense courtship lasted for about two hours and during one of the male's mounts of the female, he mated with her.

The bond between a male and his mate remains very strong after mating, and it was wonderful watching the birth of baby hopping-mice and the behaviour of the parents. In one case, which was typical, the pair had been resting, feeding and nesting together as usual although, a day or two before the birth, the female spent more time nest-building than the male. As normal, the female went into labour during the daytime and the final contractions were obvious because she hollowed her back, threw back her head, stretched out her forelimbs and lowered her vaginal region while her flanks heaved. Contractions occurred at about two-second intervals and, after three or four, the female adopted a bipedal rounded posture and vigorously licked her vaginal area. After a short interval, further contractions occurred and the head of a baby appeared. When that happened, the female again adopted the rounded bipedal posture, with her tail between her legs, and then grasped the head of the baby with her forepaws and gently guided it into the world. The newborn baby, or neonate, was born free of membranes, but it was nevertheless licked all over while the placenta was delivered. The neonate was then placed on the floor of the nest while the mother pulled the placenta away from her vagina. In this instance, the mother ate the placenta and then lifted the neonate while its umbilical cord was eaten and then nipped off close to the neonate's skin. Throughout all this, the male stayed very close to his mate. In other cases, it was the male who ate the placenta and umbilical cord, and sometimes the male took a half-eaten placenta from the female and finished eating it. Hopping-mice usually have three young per litter, and it is usually the male who takes care of the first-born while the others are being born. The position adopted by the mother prior to delivering each baby, and the prompt eating of the placenta, reduces soiling of the nest materials but, if soiling does occur, both parents spend time licking everything spotlessly clean. Then, usually, the mother pulls her young underneath her so they can suck. In one case, the father had a different idea! He gently picked up a baby in his mouth and, while partly supporting it with his forepaws, he walked or hopped across the open-cage on his hindlimbs and then put the baby down in the other nest-box. The mother did not approve of this, so she fetched her baby back in the same manner. But the male stole another baby and again the mother had to retrieve it. This went on for a whole hour but eventually the male gave up and the mother settled down to suckle all three babies.

From the moment they are born, babies are cared for by both parents – the males doing everything except suckling the young – and the bond between the parents remains very strong. During the night following the birth of the young, there is a post-partum oestrus and another mating, and the female's second pregnancy coincides with lactation that stops only a few days before the second litter is born. The interval between successive births is thirty-two to thirty-four days.

Baby Spinifex Hopping-mice are born naked, blind and deaf and, for the first five days of their lives, they can make only feeble movements with their limbs, and they seem to respond only to touch and perhaps scents. The first behaviour patterns that were not present at birth begin to appear on Day 6 and this initiates a fifteen-day period of rapid growth and elongation of the hindlimbs, during which the young learn to walk on all fours, adopt some of the adult postures, sniff at things and feel things with their whiskers. Their ears open sometime between Days 13 and 18, the pinnae of their ears begin to unfold and move in response to sounds, and they begin to be startled by sudden noises. Their eyes begin to open on Day 19 and are fully open by Day 21. During the next ten to eleven days, patterns of social behaviour develop rapidly and siblings spend a lot of time jumping over each other, pushing under each other, huddling in close contact, grooming each other, mounting (without intromission) and following each other. They sniff each other's noses, necks, perineal regions (between the hindlegs) and tails. All of their social behaviour is amicable. This period of socialization ends when the young are weaned – when they are about thirty to thirty-four days old. From weaning onwards, the behaviour of the young becomes increasingly adult-like. Amicable behaviour persists between siblings and between all members of their group. However, as the young reach sexual maturity, encounters with unfamiliar individuals elicit aggressive and repelling behaviour, just as in adults.

I never observed any aggressive behaviour at all between siblings and when successive litters were born, older siblings participated in caring for the young. I also noted that, if two females in a group had young, they cared for each other's young even to the extent of suckling them – which meant that, potentially, lactating females could safely leave their young to go out and forage for food. In captivity, the suckling of other females' young happened in groups of unrelated females, but it would be most interesting to know if it is only closely related females who suckle each other's young in the wild.

The members of groups of Spinifex Hopping-mice co-operated with each other in other ways too. They often dug burrows in tandem with the one in front doing the digging and the one or more behind kicking the loose sand back out of the way. Furthermore, groups shared a communal nest-chamber and collaborated to block the tunnels leading into the chamber when it was advantageous to raise the humidity as a means of reducing evaporative water loss from their skins and respiratory tracts. Also, by huddling together in a compact mass when it was cool, they conserved heat and were able to reduce their metabolic rates. These energy savings must be crucial in times of drought and food shortage. Yet another advantage of living in groups as these mice do, is that strong bonds between males and females form even in the absence of breeding during periods of drought, and therefore males and females can be ready for breeding as soon as food becomes available again after rains or floods – they do not have to waste time looking for mates and forming bonds. This ability to maximize reproduction when the carrying capacity of the habitat suddenly increases is known as an r-strategy and it is a characteristic of animals in habitats whose carrying capacity is subject to large and very unpredictable fluctuations.

Desert Mouse dramas

Desert Mice do not look like other native mice from Australian deserts. They are typically mouse-like in form and have chestnut-brown fur with long, darker guard hairs. The eyes are only moderately large and have a pale orange ring around them. The social behaviour of the Desert Mice from the Johnson's Bore oasis habitat was very different from that of the desert-dwelling Spinifex Hopping-mice and I believe there are two main reasons for this – they did not have to contend with intolerable conditions above ground during the day, and they did not have to contend with enormous and very unpredictable fluctuations in the carrying capacity of their habitat. The dense sedge that grew on the moist sands buffered the nests and runways of these mice against the extremes of temperature that occurred out in the open, entrapped humid air over the damp sand, and provided a constant supply of food and water in the form of succulent stems, flowers and seeds. Consequently, the number of Desert Mice living in the oasis was undoubtedly as great as the habitat could support, and any advantages of living in large groups were offset by the advantages of living alone (except at the time of mating and when females were suckling young) and defending a territory just large enough to supply all the needs of the territory-holder. This is known as a K-strategy.

In captivity, in cages furnished with sedge and other vegetation, encounters between two adult males were characterized by the absence of any amicable behaviour! Encounters were of three sorts. Sometimes they were highly agonistic with one male attacking, pursuing and fighting the other, and I had to separate these males because I did not want them to be seriously wounded. Sometimes the encounters were moderately agonistic with one male aggressive and the other non-aggressive, but fighting-in-contact was infrequent and injuries did not occur. However, each male defended his own nest-box, and they usually utilized the open-cage at different times – the aggressive male spending more time in the open-cage than the non-aggressive one. Sometimes (but less often) the 'encounters' were characterized by continuous avoidance. The males occupied separate nest-boxes, and they did not attempt to enter each other's nest-box so there was no nest-defence, and they utilized the open-cage at different times.

Encounters between two adult females were of two kinds. One was entirely agonistic at first but, after four days, there were some contacts for sniffing each other although repelling behaviour was more frequent, with one female being the aggressor and the other non-aggressive. Each female defended her own nest-box. In the second kind of encounter, the two females avoided each other completely except on Day 5 when both occupied the open-cage together for seven minutes and one was attacked and pursued by the other.

Three encounters between an adult male and an adult female that resulted in breeding either began with several days of complete avoidance or they began with agonistic behaviour during which the male was invariably the aggressor and the female invariably non-aggressive. The most frequent elements of this agonistic behaviour were attack, pursuit and nest-defence, but the amount of agonistic behaviour depended on how much time was spent together in the open-cage. In one case, a few contacts for investigation occurred on the second day but after that, until the female came into oestrus, the relationship was again agonistic but this time it was the female who was the aggressor. When the females came into oestrus, sexual behaviour began within two hours of the onset of darkness. In each encounter, the male investigated sites in the open cage where the female had urinated, and this elicited his sexual arousal. He sniffed the female, followed her and often mounted her and the relationship was entirely amicable although the female rejected the male by backing away and climbing over obstacles. This phase lasted about twenty minutes, and then the female displayed the mating posture (lordosis) either in front of the male or while he was mounting her. Then the male thrust his nose deep into the female's neck and gripped her flanks very tightly with his front legs. This appeared to force the female to adopt a more pronounced lordosis posture with her back hollowed, hindquarters raised, head up, ears back and eyes wide open. The female often made scratchy-squeaking vocalizations similar to those made by these mice when they are distressed. Then the male achieved intromission and copulation after which the female usually struggled and dislodged him. Then both animals groomed themselves. After that, if the male attempted further mountings, he was attacked and chased back to his nest-box, and the relationship became entirely agonistic with the amount of aggression depending on the amount of time both animals were in the open-cage at the same time.

The relationship remained agonistic throughout the female's pregnancy. Births always occurred in the light period or late in the dark period. The young were born in the female's nest, or in a new nest built by the female if the old nest was still occupied by a previous litter. There was a brief post-partum mating and after it, the female attacked the male and drove him into his nest-box and then, throughout parenthood, the relationship remained agonistic and, again, the amount of aggressive behaviour depended on how much time the two animals spent in the open-cage at the same time.

It was also interesting that, although young Desert Mice were amicable at first, as soon as they were weaned, the mother-offspring bonds began to break down as a consequence of aggressive acts by the mother, the young began to be aggressive towards their fathers and, although relationships between siblings were mostly amicable, some agonistic behaviour occurred between them, especially during the dark period.

The social behaviour of other conilurine rodents

Detailed observations of Silky Mice from the semi-desert habitat showed that, like Spinifex Hopping-mice, they too lived in amicable groups, but there were some differences and field-observations by my friend, Clive Crouch, revealed that these mice lived in colonies with many burrows and, while groups of females always returned to the same burrow every day (whether or not they had young in that burrow) the males occupied different burrows with different female occupants, on consecutive days. The home-ranges of the males were variable. Usually, the males ranged over restricted areas and visited only three or four different burrows, but some males ranged very widely (probably over the entire colony) and therefore had access to about twenty burrows. Clive also observed that there were periods when many new burrows were excavated, and subadults moved into these burrows. Subadult females were found singly or in twos in burrows that were visited by adult males, one at a time. Subadult males were found singly until they reached sexual maturity several months later. Adult females usually remained in their old burrows and were often pregnant or suckling young when the subadults from older litters left, and they continued to be visited by adult males. Like Spinifex Hopping-mice, Silky Mice bred throughout the year in captivity and also in the wild, but Clive believes there is a peak in reproductive activity in August, after the period of winter rainfall. The climate in the habitat of these mice is more seasonal and less unpredictable than in the deserts, and this appears to be the driving force behind the selection of the different social behaviour and reproductive strategies in these species.

In contrast to the Silky Mice, Heath Rats from the heathlands, where climate and the availability of food are markedly seasonal and fairly predictable, have one or two litters during late spring and summer, when food is maximally available, and they form pair-bonds in the non-breeding season that persist throughout mating, pregnancy and parenthood.

My most detailed investigations of social behaviour were on the four species mentioned above, but I also conducted four simple tests on other species to determine if they were group-living species or otherwise. Were they found together in nests or burrows in the field? Was the behaviour of individuals that were caged together amicable or agonistic? Did caged animals nest together or separately? And, if bonds formed, did they last throughout pregnancies and parenthood?

Using data from my own studies, and from the literature, I concluded that the desert species – Fawn Hopping-mice, Dusky Hopping-mice, Plains Mice and Sandy Inland Mice – live in large groups similar to those of Spinifex Hopping-mice. Mitchell's Hopping-mice live in smaller groups. Long-tailed Mice from wet rainforest and wet sclerophyll forests in Tasmania live in permanent pairs. And it seems that the social organization of Broad-toothed Rats from alpine habitats, where there is deep snow in winter, is intermediate between that of Heath Rats, which live in pairs, and the dispersed organization of Desert Mice.

These observations support my idea that conilurine rodents from desert habitats, where temperature and aridity can create major hazards and where the carrying capacity fluctuates enormously and very unpredictably, derive many benefits from living and co-operating together in large groups. Similarly, species from semi-desert habitats also benefit by living in groups, albeit smaller ones. In contrast to these r-strategists, species from non-desert habitats, where fluctuations in carrying capacity are either small and predictably seasonal, or absent, do better by adopting K-strategies in which the advantages of living in large groups are outweighed by the advantages of defending territories and living in pairs or singly (except during mating and while females are suckling young).

The courage, innovations and adaptability of rodents

Following the advice of Max Downes who believed that anyone studying an animal should keep some as pets, I kept a group of Plains Mice at home, as pets. I kept them in a large glass 'aquarium' furnished with sand on the floor, some dried grass and a small pile of logs. Consequently, these burrow-living mice had nowhere to burrow into, and no nest-box either. Quite undeterred, they built themselves a communal nest of grass tucked into a hollow within the pile of logs. We had the 'aquarium' in our dining room and, while we sat at the dining table, my parents and I watched the mice as they reared babies and went about their routine activities.

Some other groups of Plains Mice were housed in Perspex-sided cages in the animal house at Monash University. At the bottom of each cage there was a removable metal tray containing wood shavings and sawdust and a single nest-box. The animal house was well ventilated by air conditioners, but these blew a draught of air through the narrow gap between the Perspex side and the metal tray of one cage, and then this draught blew straight into the opening of the nest-box. What did the mice do? They built a wall of shavings and sawdust at an angle of forty-five degrees to the hole in the nest-box, and this prevented the draught from blowing into their nest!

I am always very impressed with the adaptability and innovative behaviour of mice. No matter what happens to them, they just get on with their lives as best they can. I had litters of hopping-mice born in tiny holding-cages while they were being flown from Sandringham Station to Melbourne: their mothers had built nests and were suckling and caring for their young as though nothing unusual was happening. And, I have known many rodents in Australia and elsewhere, who gave birth inside live-traps but then looked after the babies as though they were safe in their own nests in the wild.

I think rodents also have great courage. A friend and I watched a Black Rat in a feed shed who had built a nest and given birth to a litter in a forty-four-gallon bin with a loose lid that contained old sacks and other odds and ends. When we took the lid off, the rat panicked and leapt out of the bin, and took cover. We stood aside and watched. After a few minutes, the mother rat came out into the open, found her way back into the bin, and then carried the babies, one at a time, to a place she had found under cover somewhere else in the shed. Rats in feed sheds near stables are well-known pests, but it was impossible not to admire the courage of this valiant little mother.

It seems to me that rodents never (or only very rarely) 'throw in the towel' and lose the will to live. Instead, they simply make the most of whatever comes their way and just keep on going – so it is no wonder that they have thrived more or less all over the world.

6

MEANDERINGS WITH A MONGOOSE
AND OTHER RAINFOREST ANIMALS IN NIGERIA

David

In September 1967, when I was working on the ethogram of the Spinifex
Hopping-mouse, my supervisor thought I should read a paper by D. C.
D. Happold describing the behaviour of the Lesser Egyptian Jerboa in the
Sudan. Jerboas are bipedal desert rodents that are ecologically equivalent to
the Australian hopping-mice. So my supervisor requested a reprint from
the author and, when it arrived, it was accompanied by a brief note asking
for information about the Marsupial Jerboa. My supervisor showed me this
note and said, "I think you should answer this." So I did. It was the most
important letter I have ever written because it began an almost four-year-
long correspondence during which David Happold and I came to know
each other very well. In 1971, David spent six months of sabbatical leave
at Monash University while I was finishing my Ph.D., and we soon found
out that we wanted to spend the rest of our lives together. We were married
in October 1971 but, all too soon, the idyllic time we had had together in
Australia came to an end. David had to return to Nigeria (where he was
now living) at the end of December to resume his teaching, and I had to
stay behind at Monash until I had completed some crucial observations for
my Ph.D. We had Christmas with my family – and then he flew away to
Nigeria and it felt as though I had lost him forever.

58

Journey to Nigeria

I was not able to follow David to Nigeria until May, and those five months were the worst in my life. The rule at Monash was that Ph.D. students had to submit their theses before leaving the university. This was mainly so that the writing of the theses could be followed closely by the supervisors. However, my supervisor had sabbatical leave in 1972 and went to Munich to work with Nobel laureate Konrad Lorenz, so it was agreed that I could write my theses in Nigeria where David could keep an eye on my progress.

David had a special request. He had found it very frustrating trying to teach Nigerian students about reptiles because they refused go anywhere near any Nigerian examples. He thought they might feel more at ease with a docile Blotched Blue-tongued Lizard from Australia, so he took one with him when he returned to Nigeria and he was a great success. So David asked me to bring one or two females to be company for Dominic as he was called. I actually acquired three females, and then each gave birth to six babies, but I obtained a permit to take all of them out of Australia and I knew they would be very welcome at the Ibadan Zoo which was attached to the Zoology Department at the University of Ibadan. The only problem was that I could not take the lizards with me because I was travelling via South Africa where the lizards were not permitted, even in transit, so they had to travel via Rome, along with some unaccompanied baggage, and therefore they had to leave a day before I did, and they arrived in Nigeria several hours before I did.

I should have had very mixed feelings about leaving 'Willow Hill' but my longing to be with David obliterated everything else. However, I did not sleep very well the night before leaving and I crept out of the house and went to find my beloved horse Camie. He stood beside me with his head resting on my shoulder and getting heavier and heavier as he dropped off to sleep. I didn't know if I would ever see him again.

I set off from Tullamarine Airport near Melbourne and, after a brief stop in Johannesburg in South Africa, I got into the plane that was going to take me to Kinshasa in the Congo and then Lagos in Nigeria. At Kinshasa Airport, I had my first experience of tropical heat and humidity, and it was like being in a sauna except that it smelled of tropical Africa – moist and dank. I was only briefly in transit and my first experience of African officials was uneventful and pleasant. Then I was off again, flying over dense rainforest, patches of savanna, wide rivers, subsistence farms and small villages.

Culture shock!

And then we reached Lagos – endless square miles of rusted corrugated-iron roofs and narrow roads. I was first out of the plane – and there was David! He had been allowed to come through customs, and so he was able to help me through the formalities of immigration. I do not remember much about all that – I only remember the perfect joy of being with David again.

I had my cabin luggage with me – the essential things I would need for writing my thesis – but the case that had gone into the hold was missing (and it was not sent on to Ibadan for several days). Furthermore, no one knew anything about my unaccompanied baggage which should have arrived several days ago, and no one cared. But *everybody* knew about the lizards! They should have been kept at the international airport until we collected them but, instead, arrangements had already been made for them to be flown to Ibadan in the morning, on the first possible plane. In the meantime, they had been moved to the domestic airport and had been locked into a shed where they were meant to remain overnight. But yes – someone would take us to the domestic airport, and someone would unlock the shed and hand over the lizards, even though it was now late in the afternoon and most of the staff had already gone home. I was *most* impressed with the efficiency generated by fear of those lizards, but *not* impressed at all with the efficiency generated by a new arrival's urgent need for the clothes in her unaccompanied baggage. They said it would be delivered to Ibadan, but many months passed before that actually happened.

We got the lizards and took them and my cabin luggage to David's Opel car. Unfortunately, its brakes had started to give trouble during David's journey from Ibadan so, instead of driving back immediately, David had arranged for us to spend a few days with some friends while the brakes were fixed. This meant driving through Lagos – and culture shock began to set in. It was brown and black everywhere. Traffic chaos. Every car horn blaring. Piles of rubbish where dwarf goats fossicked for orange peel or anything else they could eat. Horrible open, deep ditches at the edges of the roads. Bridges over polluted waterways that were pale grey and choked with rotting rubbish. In contrast, there were dense crowds of Yoruba people either in colourful traditional clothing or in immaculate suits with white shirts and ties. And yet none of this really sank in. I was with David again and that was all that mattered.

We had to stop at a service station on the way to the friends' house and, while David was getting petrol and making arrangements for the brakes to be fixed, I opened the lizard-box to check how the lizards were coping. They were all well, and willing to eat some fresh banana. Unfortunately, someone spotted them and I was immediately surrounded by children and adults, all talking at the tops of their voices and trying to decide whether to stay and look, or bolt while they still could. I had no idea that those lizards would cause such a stir and David had to come over quickly to help me get them out of sight.

After two nights with the friends, we set off – with repaired brakes – to drive 125 kilometres north to Ibadan, and the culture shocking continued. When Lagos was behind us, the road opened out and we travelled though patches of creeper-lined rainforest, swampy areas with oil-palms, subsistence gardens with cassava, yams, maize, tomatoes, mango, orange and avocado trees, and small towns or villages with two-storied mud-brick houses packed closely together without any greenery around them. The road carried a lot of traffic – mostly what David called 'mammy wagons' that were dilapidated trucks, overloaded with people or goods and belching black smoke. Every mammy wagon had a wooden board with a slogan above its front windscreen. I was amused that the slogan of one that was belching particularly dense clouds of black smoke was 'No Smoking'. The slogan of another that was missing on several cylinders and only just getting up the hills was 'Struggling Man'. One, whose driver was overtaking on a crest with a blind curve, had chosen 'God First' and another, who had wrapped his wagon around a tree with an impact that made it almost unrecognisable, had chosen 'No Condition Is Permanent'.

There were people wandering along the roadside, usually with heavy loads balanced on their heads or pushing handcarts. There were dwarf goats of various colours keeping themselves cleverly out of the way of vehicles and, in contrast, sheep, chickens and dogs making suicidal dashes in front of the vehicles. There were local men struggling to drag kicking, rope-festooned cattle somewhere they did not want to go and, in contrast, there were a few white-robed nomadic Fulani herdsmen from the north guiding their calm, dignified cattle through all the chaos, with nothing more than their quiet voices and a gentle touch with a long stave. By the time we reached Ibadan, words had started to fail me and I actually found it difficult to talk for several months. Culture shock is a very real affliction and, at this time, I had yet to be exposed to all the really exciting things about Nigeria – its vibrant culture, literature, traditional music, dancing, arts and crafts and, of course, its animals.

Ibadan! Again, black and brown were the dominant colours, but the buildings were an incongruous mix of modern skyscrapers and old mud-brick houses. The noise was unbelievable! Blue-and-yellow taxis, with smooth tyres and battered sides, dodged through the traffic with their horns hooting non-stop. Every roadside shop or stall had a radio blaring out Yoruba pop music, singing or drumming, and there were always children there, moving in time to the music. Boys rolled tyres or used sticks to bowl hoops, and if they saw us in the car, they shouted "Oyinbo! Oyinbo!" as though they had never seen a white person before. We passed many markets including a goat market where I saw long-legged, elegant, chestnut-brown goats from northern Nigeria for the first time, and there was a market where nomadic Fulanis sold their majestic, white, long-horned zebu cattle. There were other markets that specialized in earthenware pots, or cane furniture and baskets, or cages full of chickens. The roads were lined with dilapidated sheds in which there was a multitude of things for sale – powdered milk, sugar, bread, clothing, cloth, tins, buckets, brooms . . . And there were stalls where carpenters made crude furniture and held their saws vertically with the teeth pointing away from them! There seemed to be huge billboards everywhere advertising such things as Nestlé powdered milk, toothpaste, cures for malaria, and beverages such as Coca-Cola and beer. And there were hundreds of small signs advertising such things as 'The Good Hope Wear Tayloring Centre', 'Challenge Guest House', 'Trust in God Motor Mechanic', 'Up Hill Driving School' and 'Mercy's Beauticotique'.

There was an area between Ibadan and the campus of the University of Ibadan, which was a mixture of scattered houses, patches of palms and trees, tall grasses and plots of vegetables. We started seeing signs to Ibadan Airport, invariably with the planes flying upside-down because they had been moved from one side of the road to the other in early 1972 when Nigerians started driving on the right instead of the left. And then we passed the Ibadan Airport and soon came to a high chain-link fence surrounding the University of Ibadan campus – and suddenly the atmosphere changed.

Introduction to life on the University of Ibadan campus

A wide road led up to the main gate where security guards sometimes watched from a sentry box. There were some tall and very beautiful Rosy Trumpet Trees around the main gate and, beyond the gate, there was a wide two-lane avenue with a median strip and sides planted with shady rain trees. This avenue led towards the porter's lodge and then to residential areas with attractive colonial-style houses in big gardens. Further on, there were some of the students' halls of residence and then the first of the university departmental buildings.

We turned off the main road just past one of the largest of the student halls and followed a short road that led to Kuti Hall, the block of flats where David lived on the ground floor. The outside was scruffy but, inside, David had everything clean, tidy and ready to welcome me. The flat was small but it had a kitchen, a dining-room with a large table and French windows leading into the garden, a bathroom with a bath, hot-water heater and flushing toilet, and a spare room with a spare bed for guests. The floors were concrete polished with cardinal red polish, and the brick walls were smoothly plastered with concrete and painted cream. The flat had electrical power but it often went out, especially during the wet season, and I learned that NEPA stood for 'No Electrical Power Again' (not 'National Electrical Power Authority'). There were also pipes and taps for running water, but it didn't always run, so we kept the bath full of water and used this when necessary. Outside his flat, David had paved a patio and planted sweet-scented jasmine and other climbers around it, and then there was a garden, dominated by a large flame tree and a frangipani, with colourful tropical shrubs around the boundaries and a vegetable garden at the bottom.

Almost as soon as we arrived at Kuti Hall, I was introduced to David's steward, Moses Okpeke, and his wife, 'Mrs Moses', and children. Moses had worked for David for five years (just about a record) and he knew just what David expected of him. I hoped my arrival would not rock that boat, so I set out to be very kind and friendly – as I would have done had he been a fellow Australian.

The night fell with equatorial rapidity and I listened to the sounds. Lions in the zoo roared and coughed. Toads croaked. Franquet's Epauletted Fruit Bats called from the trees and sounded like cracked bells, and the calls of smaller species of fruit bats sounded like tinkling bells. Radios were turned on loudly, and 'Lillibolero' marked the beginning of the BBC news. And there was the steady thud, thud, thud of women pounding yams.

My life at Kuti Hall settled down to a steady routine. We often ate breakfast on the patio, and after breakfast David walked to his office in time for his first lectures, which were usually at eight o'clock. My top priority was to finish my thesis and I worked on the dining-room table where I could look out to the garden and listen to birds such as the Common Bulbul who cried "Quick doctor, quick" and Red-eyed Doves whose call was "coo-coo, cu-KOO-coo-coo. I am a red-eyed dove". There were also brilliantly coloured sunbirds, Kurrichane Thrushes, Red-cheeked Cordon Bleu Finches and many other birds in the garden.

During those early days, David took me into the Zoology Department to see his office and various collections, and to meet the other members of staff. I also saw his animal-house where he had cages with various native mice such as Tullberg's Soft-furred Mice, Edward's Swamp Rats and Rusty-bellied Brush-furred Rats from the rainforest, and Tiny Pygmy Mice, Dalton's Soft-furred Mice, Zebra Grass Mice, Kemp's Gerbils and Slender Taterils from the savanna. The Zebra Grass Mice were exquisite. And we often went to the zoo that was run by the Zoology Department under the management of Bob Golding. The first animal David wanted me to see there was Ho-Ho, a Two-spotted Palm Civet that he had raised and kept as a pet in his flat until she grew-up and became untrustworthy. She was very beautiful but not very friendly. Quite the nicest animal was a Rock Hyrax that loved to be petted. As well as Ho-Ho, there was a Cusimanse Mongoose, several Large-spotted Genets and African Civets, all of which belong to the Viverridae – that strange family midway between the Felidae (cats) and the Canidae (dogs) that contains both cat-like and dog-like representatives. There were also many sorts of birds, crocodiles and large mammals such as African Elephants, Striped Hyaenas who seemed very friendly and always came to the fence (but I was warned not to touch them), Lions who were anything but friendly, Patas Monkeys and other monkeys, and Western Gorillas.

The Blue-tongued Lizards were also kept at the zoo and it was lucky that Bob Golding was a reptile expert because the three females and the babies did not adjust to the change in climate from the cold of Melbourne in May, which had triggered their hibernation, to the heat of Ibadan. But Bob believed he could save them by exposing them to extra heat, and that worked. There was never any trouble with Dominic who had come to Nigeria from mid-summer in Melbourne. He was a great success: the zoology students, who are traditionally afraid of lizards, would examine and handle him.

This reminds me of Moses and a chameleon! Nigerians are especially fearful of chameleons. I did not know this when I found my first chameleon and I was so fascinated by its stealthy creeping movements and independently swivelling eyes that I brought it into the flat and let it go to wander about at will. I came in one day to find Moses polishing the concrete floor with his polishing rags attached to his feet, as usual. He used to do a sort of dance while sliding the cloths over the floor but, on this occasion, his eyes were absolutely fixed on a corner of the ceiling and he kept bumping into the furniture as a result. The chameleon was up in that corner!

We did not go outside the campus much, except for shopping trips into the centre of Ibadan, but of course David took me to Gambari Forest to participate in his weekly trapping sessions. He was conducting a long-term demographic study of the small mammals in this almost pristine rainforest that had tall trees with buttress roots, strangler figs, lianas and all sorts of ferns, as well as other forbs growing on the ground. Rotting logs had brightly coloured fungi growing on them and, if the ground was very moist, there were often clouds of butterflies sucking up the moisture. David knew almost every animal on his trapping-grid, and he would greet his trapped animals like old friends. Particularly nice were the gentle, velvet-furred Edward's Swamp Rats.

I arrived at the beginning of the first wet season. In Ibadan, there are two wet seasons with a short 'dry' period of much less rain between them, and a long dry period. Nigeria is wettest in the south and driest in the north. Just north of the coast, there is the rainforest, then there are three bands of increasingly drier savannas and then the edge of the Sahara Desert. Ibadan is located on the boundary between the rainforest and the first of the savanna zones. The wet season in Ibadan was a new experience for me. I had never experienced such dramatic thunderstorms or such torrential rain! And it was so hot and humid that one perspired all the time and mould was a real problem. It grew on anything made of leather and all our photographic slides had to be kept in glass desiccators containing silica gel. However, because Nigeria was considered unhealthy for expatriates, David's contract included three months of annual leave in England during July, August and September, so we always missed most of the wet seasons.

My first full academic year in Nigeria

David took me to some interesting places during the academic year of 1972-73. On one occasion we went east to the University of Ife at Ile-Ife, where research was being done on the large African Straw-coloured Fruit Bats, which roosted in vast numbers in trees on that campus. We used to see them flying over the University of Ibadan on their long foraging flights in search of fruiting rainforest trees. We also took a few weeks off and went north into the Derived Savanna and Guinea Savanna zones south of the Niger River to do some trapping for rodents and mist-netting for bats. We drove about 300 kilometres to Kabba where we camped in the grounds of an agricultural station and, after that, we stayed at a school in Egbe, a hospital rest-house at Omu-Aran, and a government rest-house at Ejigbo. While we were staying at Ejigbo, we went to a traditional night market where there were lots of tables or rugs with all sorts of things for sale, and everything was illuminated by kerosene lamps. It was nice that we could wander through such markets without any concerns. In fact, we were always made to feel very welcome everywhere in Nigeria.

On that trip, and on most of our journeys north of Ibadan, we often saw Fulanis with herds of their white, long-horned zebu cattle. The cattle Fulani people are nomadic cattle-herders who wander north and south across the savannas, following the rains in search of pastures for their cattle. They are tall, lean, very dignified men who wear long whitish robes and usually walk with their arms bent over a long wooden stave on their shoulders. We knew that the herdsmen were unable to carry much water with them, so we always carried cans of water in our car so we could give them a drink. It was wonderful witnessing the rapport between these herdsmen and their cattle, and I often wished I could walk with them on one of their long migrations.

Rescuing a nestling sunbird

One day in mid-January, when I came back to Kuti Hall from a visit to Kingsway (a small supermarket on campus) I found Moses's children waiting to show me something that one of them was holding in her hand. It was a nestling Olive-backed Sunbird and it was lying on its back very close to death. I told the children that it was very sick, and I persuaded them to let me look after it. Luckily, we had some honey, so I made some honey-water and dipped the tiny bird's long beak into it. The tongue moved in and out a few times, and some of the honey-water must have been swallowed. I repeated this every hour during the day and well into the night and, by the time we went to bed, the bird had regained enough strength to right itself and had started to look around alertly. It was far too young to fly or even hop around, so I made a nest of cloth in the bottom of a large wicker laundry-basket and settled it in that. I did not expect it to be alive in the morning, but it was and it began to make 'cheep cheep' calls. That day it fed well by lapping honey-water from a spoon, and by midday it had begun to hop around and could perch on a stick wedged across the basket. I usually had a siesta after lunch, but I had the basket at the foot of the bed, near the open window, so that the bird's cheeping for food would wake me up when it was hungry. One afternoon, I was woken up by the baby suddenly making loud, persistent cheeping, and then an adult female sunbird flew into the bedroom and down to the rim of the basket. I stayed very still and watched. The female made a call and flew down to the baby, then back to the rim several times as though it were trying to entice the baby to follow it. The baby did try to follow, but the steep sides of the basket were impossible for it to climb, so I put the baby on the windowsill and the female flew to and from it, again enticing it to follow. The baby tried to hop and flutter after the female, so I took it into the garden and let it perch on the branch of a hibiscus shrub. Throughout that afternoon, the baby was fed insects by the adult female and also by an adult male, and I also gave it honey-water once every hour until dusk when I brought it into the house again and settled it in the basket for the night. Next morning, I heard the adults calling from the garden at about eight o'clock, so I gave the baby a feed of honey-water and then returned it to the hibiscus. Throughout that day, the adults fed it and, by the afternoon, it had learned how to hop from one twig to another. However, on one occasion, it lost its balance and found itself hanging helplessly upside-down. The female landed on the twig and attempted to push the baby up with her bill and forehead. The first attempt was unsuccessful and she flew away, but less than a minute later she came back and hung herself upside-down beside the baby. In this position, she was able to push it into an upright position! At dusk, I again brought the baby in for the night, and put it out again next morning. That day, the baby managed to get itself right down to the bottom of the garden by nightfall, but I found it and brought it in again. By the next day, it was moving around the garden quite a lot and by dusk it had moved out of the garden and I did not see it again. It had been an interesting five days, but it raised an interesting question – were the two adults who fed the baby its parents, or were the baby's calls enough to trigger parental behaviour in unrelated adults? According to ornithologist Hilary Fry, the latter is known to happen.

Moving to No. 1 Ebrohemi Road

When we returned to Nigeria in mid-October after our annual leave, I was three months pregnant and this meant that David now had almost enough points to qualify for a house to live in and he put in an application. We were very lucky. Although we would not have the requisite number of points until our baby was born, there was a house (on a large block of land) that was in such a dreadful state that no one else would have it, and it was offered to us. David said we would take it provided it was repainted and rewired, and that the Estate Department would send a team of gardeners to clear the jungle of unwanted vegetation and rubbish that had accumulated where there was meant to be a garden. And the Housing Department agreed to everything!

Clearing the block was the first thing to be done and it took a team of eight gardeners two weeks to do this. The jungle had encroached, and there were stick-trees (garden stakes that had taken root) growing everywhere. Very little was worth saving except for a large flame tree, a huge rain tree, a mango tree, a frangipani, a bougainvillea growing over the kitchen, a scraggly euphorbia, a few hedge plants (mainly red hibiscus) around the boundaries, and a magnificent avocado tree. There was also a stand of banana plants and a huge clump of giant bamboo that provided all the bamboo poles we ever needed. It was great fun having almost an acre of land to convert into a garden and eventually we planted more than 300 trees, shrubs, flowering creepers and flowers during the wet seasons. We employed a gardener to help but David and I did much of the work ourselves.

We established several beds for shrubs and a large vegetable garden and, because the soil was very infertile laterite, we brought many sacks of elephant manure from the zoo and dug this into all the beds. It immediately sprouted hundreds of pawpaw trees, but rapidly turned the pale red laterite into rich, dark soil. We planted most of our vegetables in the improved soil, but we also planted some in poor soil and then David brought his students to see the difference. They were amazed at what manure and compost could achieve because Nigerian gardeners and farmers traditionally do not do anything to improve their soils. Instead, they follow a 'slash and burn' regime and simply move onto newly cleared ground when their crops start to fail.

The house was the next thing. All the repairs, painting and rewiring had to be done by workmen from the Estate Department – we were not allowed to do *anything* ourselves. However, we were allowed to choose the colours and I made curtains from material we bought in Ibadan. The floors were concrete polished with cardinal red polish and, in lieu of skirting boards, a band of red was painted around the bottom of the walls. The painters needed a lot of supervision but, after they had been painting for several weeks, one of the men came to me and said, "Now we understand. You want this to look like a *new* house!" "Yes" I replied. And, in the end, it did – and it was our first real home!

Pet tree hyraxes

We had some interesting pets during my first year at Ibadan. Soon after we moved into 1 Ebrohemi Road in February, we heard about a family in Lagos who were about to return to England, and they needed someone to take over the care of their Western Tree Hyraxes – a pair with two juveniles. We asked Bob Golding at the zoo about these animals and he said tree hyraxes rarely survived being taken from the wild into captivity – they almost always pined away and died. In this, Western Tree Hyraxes differed from Rock Hyraxes, which adapt well to captivity and make wonderful pets. Bob thought the tree hyraxes from Lagos would not stand being moved to

Ibadan, but he had read recently that it was possible to cure pining animals by over-stimulating them and we were willing to try this. We had a 3 × 2 × 2 metre enclosure built for them in our garden, and we furnished it with branches for climbing on and a nest-box. From the time they were first put into the enclosure, I gave them very little peace. I went into the cage every hour or so, picked them up and put them down somewhere else, talked to them, stroked them, moved things around the enclosure and did everything I could think of to prevent them having time to think about their plight. It did not work with the juveniles who died after two weeks, but the day came when, instead of sitting around listlessly, the male was so frustrated by all my fussing that he attacked me and sank one of his long tusk-like teeth deep into the calf of one leg! This was a breakthrough (in more ways than one) and from then on, the behaviour of both hyraxes was normal and they began to thrive.

Tree hyraxes are somewhat like very large guinea pigs. They have short legs and their four-toed forefeet and three-toed hindfeet have soft pads underneath, and the toes have flat nails instead of claws. Furthermore, they have no tails! One cannot imagine an animal less suited to living in trees but, in fact, they are superb climbers and they can achieve – by balancing on their little feet – what other climbing mammals achieve by hanging on with sharp claws and long prehensile tails. Hyraxes are actually related to elephants, and the tooth that sank into my leg is anatomically the same as an elephant's tusk. The tree hyraxes did not make particularly attractive pets and before we went on leave in 1974, we moved them into an enclosure at the zoo where they settled down well.

Our darling mongies

The best pets we have ever had arrived in a brown paper bag, about a month before our baby was expected. They were two very tiny baby Cusimanse Mongooses, a male and a female, that had been offered to the zoo. Being carnivorous, they were easy to rear. At first, we gave them milk from an eyedropper and then we gave them minced meat and any insects we could find. The mongooses were very like puppies. They played with each other, and had furious fights with shoes, mats and the waste-paper basket. They had long, inquisitive noses that poked into every nook and cranny and soon there were no cockroaches in *our* house! Just like puppies, the mongooses could be left to run free in the house and in the garden. They came when we called them and, in the house, they used a tray full of sand that substituted for kitty-litter. They made a wide variety of calls, including a high-pitched 'weeeee weeeee weeeee' that expressed excitement and delight, and a deep growl when they were losing their battles with the shoes, mats and waste-paper baskets. We adored them!

Life with a baby, a thesis, a mongoose and Driver Ants

Our beautiful little daughter, Karen Lena, was born in April. We brought her home and I got straight back to my thesis because it *had* to be submitted by the middle of the year. Monash University stipulated that Ph.D. theses had to be professionally typed on an electric typewriter, but this was not possible in Nigeria in those days, so we had brought a manual typewriter back from England and fitted it with an electric typewriter ribbon, and I did the typing myself. I worked on the dining table and had Karen Lena's bassinet in a stand beside me.

At first, our young mongooses either romped around the house, or outside in the garden. There was always the risk that someone would come into the garden and catch them for 'chop chop' but the mongooses were very wary of strangers and either ran under the house, or hid in the middle of a large, thorny cactus-like euphorbia in the back garden. But there came a weekend when I had a serious attack of malaria. David stayed by my bed and no one kept an eye on the garden. It was terribly sad. Someone *did* come into the garden and cut down the euphorbia, and the male mongoose was taken. The female had run under the house and she escaped, but she was very traumatized by the loss of her brother and, from then on, she almost always stayed close to my feet. While I was typing, she only ever left my feet to make a quick dash to the 'kitty-litter' tray – or to sneak into the bassinet where she curled up with Karen Lena! I always removed her as soon as I discovered her there, but in fact we trusted her implicitly.

Let me finish the mongoose story. We had three months of annual leave from July to September, so we tried to introduce our Mongie to the male mongoose at the zoo, thinking that she would be happy there. However, the male was very territorial and he attacked her. Since we had now moved the tree hyraxes to the zoo, their enclosure was vacant, so we got little Mongie used to living in it, and we paid Moses to stay on and look after her while we were away. We wondered how she would react when we came back to Nigeria and, as we walked into the garden, I said to David, "Let's just open the dóor and see what happens. If she runs away, so be it." We came within sight of the enclosure, and Mongie saw us and started racing round and round the enclosure. I opened the door and she flew onto my shoulder and ran round and round my neck, 'de-lousing' me with her teeth, and squealing "weeeeeeeee weeeeeeeeeeeee, weeeeeeeeeeeeeeeeee" with ecstatic rapture! Then she leapt from my shoulder to David's and repeated the performance. We have never been greeted by any animal – not even by any of our dogs – with *quite* such exuberant delight! Beyond any shadow of doubt, that little mongoose loved us.

After that, Mongie lived in the house again most of the time, but sometimes she was put back into the enclosure. Some time later, we tried again to pair her up with the male from the zoo, but this time we cleaned everything out of our enclosure, hosed it down and furnished it with new logs and branches so it was neutral territory when we put both mongooses in it at the same time. It worked. The two became friendly – at least to some extent. However, Mongie always preferred to play with us, and that made the male very jealous. He expressed his jealousy one day by attacking and biting David – so we had to be careful of him after that. When we went on leave in 1974, we moved both mongooses back to the large enclosure at the zoo and there they lived together for several years – but they never produced any little mongies, which was very sad.

Not long after Karen Lena was born and while we were getting Mongie used to living in the enclosure, our garden was invaded by Driver Ants. I had seen columns of these terrible black ants in Gambari Forest where, occasionally, they attacked mice that David had trapped, and then he would find nothing but a skeleton in the trap. Colonies of Driver Ants are the largest colonies of any social insect – colonies of fifty million have been reported! The workers are about half a centimetre long, but the soldiers are about 1.5 centimetres long and are equipped with powerful biting mandibles capable of cutting out pieces of flesh. Driver Ants are nomadic and move to new foraging grounds in long columns guarded by the fierce soldiers. Columns of solidly packed ants are usually about four centimetres wide and are often more than a hundred metres long. Periodically, the columns break up and the ants spread out to forage, and to attack and devour anything edible that they come across. Driver Ants in West Africa have a fearful reputation. There are records from the old days of mothers finding the skeletons of their babies in their cots after a nocturnal raid! I do not know what made me wake up the night of our raid, but I put the light on and found a stream of the ants just starting to come into our bedroom through a tiny gap in the flywire netting over the window. Then I looked out over the moonlit front garden and found it literally shimmering with black Driver Ants! Karen Lena was safe because the legs of her cot always sat in wide tubs of water – but what about Mongie locked outside in the enclosure from which she could not escape? David raced to Moses's quarters and enlisted his help, and I put on gumboots and raced through the ants to rescue little Mongie. The ants were already in her long, thick fur but they had not yet started to bite, and soon she was safely inside the house. For the rest of the night, Moses, David and I fought those ants with everything we had. We put insecticide around the bedroom and elsewhere in the house, but we soon used all we had. Then we attacked the invaders with kerosene, and we fought them off with water squirted from hoses. But it was the sun that finally dispersed them, and where they went no one knows.

About two and a half months after Karen Lena was born, and despite an abscess the size of a hen's egg at the base of my right thumb that stopped me typing with my right hand for many weeks, I managed to finish typing my thesis in time to take it to England in July and post it to Monash.

Life in Nigeria without a thesis

After our annual leave in 1973, which we spent in Austria and England, it was good to come back to Nigeria, free of my thesis – although I still had to write papers on the results of my research.

Toads in the garden

I spent a lot of time in the garden that year and one of my projects was to create a fernery with a pond under the Flame Tree that grew in front of our bedroom window. I heaped good soil around the base of the tree and planted it with ferns transplanted from Gambari Forest, and I stocked the pond with guppy fish to eat any mosquito wrigglers. It was a great success except that the pond attracted large African Toads – and those toads croak. Oh, how they croak! One night there were four of them – a tenor, a baritone, a bass and a basso profundo – and we could not sleep for the noise. Something had to be done, so I got into gumboots, lit a hurricane lantern, put on some gloves and went out to remove them. I caught the tenor, carried it to the far side of the garden and put it down at the back of the stand of bamboos. Then I hurried back to the pond and caught the basso profundo. I let it go at the same place behind the bamboo and then went back to catch the baritone. As I was taking it towards the bamboo, I was surprised to see a toad hopping past in the opposite direction. I released the baritone and went back to the pond to catch the bass – but there was now a tenor as well as the bass croaking in the pond. I caught the bass and headed off towards the bamboo and was passed by another toad hopping in the opposite direction and, by the time I got back to the pond there was a basso profundo again as well as the tenor! Eventually the penny dropped – those toads were hopping back to my pond as fast as I removed them!

Rearing baby genets

About the time of Karen Lena's first birthday in April 1974, we acquired three very tiny Large-spotted Genet kittens. They had been collected from Gambari Forest by hunters who hoped to sell them to the zoo, but the zoo sent them on to us. They were a female and two males. Genets are members of the family Viverridae and in many ways they resemble slender, short-furred domestic cats. But they have banded tails and a pattern of dark brown blotches and lines on the pale brown fur of their heads and bodies. The babies were easy to feed on milk from an eyedropper and then finely chopped meat dipped in egg-yolk, and banana. They began to chase and eat insects when they had been with us for about a month. Karen Lena adored them and loved to feed them even though she was only a year old herself. At first, they were fairly aggressive and spat at us like angry cats and sometimes tried to bite, but they soon became very friendly, followed us around, purred like cats when being stroked, and loved to sit in our laps. They made a very cat-like but drawn-out 'miauuu' when they were hungry and a 'chip chip' call if they became separated from each other or from us. They also made a low, drawn-out dog-like growl when they were feeling aggressive. The genets could not be house-trained and, by the time they were about a month old, they had play fights with each other. They also sometimes spat at Moses and me and occasionally bit us. However, we lightly smacked their noses when this happened and they were immediately friendly again. But the genets were very different to the mongooses, which are very social, group-living animals. As the genets grew older, they became less friendly and more independent. They were free to go outside and began to hunt their own food and eventually they became completely independent and returned to the wild.

A tale of a thicket rat

David continued trapping the rodents in Gambari Forest throughout all our time in Nigeria, but he was also interested in small mammals from other habitats – one of which was disturbed rainforest surrounding Aroro, a small village about fifteen kilometres north-east of Ibadan. David devised a clever means of finding out what lived in this forest. He visited the local primary school and offered to pay ten kobo for every mouse or shrew that any child brought in, provided it was an adult and uninjured. Nigerian children often catch small mammals for food, and they are experts at making live-traps from tins, the lids of tins, bits of wood or sticks, and elastic rubber straps cut from old bicycle tubes. When David went back to the school two days later, dozens of children, from all classes, poured out of school and lined up to present him with their catches and to receive their cash. On one occasion, I went with him and it was great to see what a thrill this event gave the children. They were *so* excited! But I witnessed something that could have been very sad. I watched as one very young boy came to David with a mouse in his hand. It was a tiny baby with its eyes barely open, and it was cold and moribund. David could not accept such small animals: he just had to be strict or he would have been inundated with animals that were of no use to him, and most of the lads who had brought unacceptable animals laughed it off when they were turned away – but not this little fellow. He looked pathetically miserable and I concluded that he was desperate to earn ten kobo for some very important reason. So, I asked him if I could look at his mouse and then I said, "I think you have a very special mouse there, and I will pay twenty kobo for it if you will sell it to me." He was overjoyed! And I had a baby mouse that was almost dead and far too small to be identified. However, it lived and it was indeed special! It was a Shining Thicket Rat, a species that David rarely trapped because it is a climbing mouse that usually forages in the low branches of shrubs and thickets – not on the ground. I warmed it up and fed it on milk from an eyedropper until it was old enough to start eating seeds, oatmeal, groundnuts (peanuts) and other solids. It was the only small mouse I ever had that was *completely* tame. It would try to follow me, and it would come running to take titbits from my fingers. We often let it run around the house because we knew we could always catch it again. It was also the only Shining Thicket Rat that David ever photographed, and there is a lovely photograph of it in David's book, *The Mammals of Nigeria*, which was published in 1987.

Our last months in Nigeria

In November 1974, our son, Jonathan Richard, was born. We were delighted to have our 'pigeon pair' and it was not long before we began to refer to them as the 'Haplets'.

By the beginning of 1975, David and I were starting to realize that things were changing in Nigeria – changing for the worse as far as we were concerned. Many of our closest friends were deciding to leave, and we were starting to feel that our children should be brought up somewhere safer and healthier, where they could receive a better education without having to attend boarding school in England. David was due for more sabbatical leave, so we spent it at the University of New England in Armidale, N.S.W., and David looked for employment in Australia. He was offered a lectureship in the Zoology Department at the Australian National University and he accepted it. We returned to Nigeria at the beginning of 1976, knowing that we had only six more months before we left, and we were determined to make the most of it.

Collecting bats in Gambari Forest Reserve and at Ipole

During those last months in Nigeria, David and I had two trips to catch fruit bats because we were planning to write a paper on 'The Fruit Bats of Western Nigeria' for the journal *Nigerian Field*. The first trip was to Gambari Forest where David had netted bats before I came to Nigeria. We stayed in the old forestry rest-house and set nets in the lower strata of the forest and in small clearings. We only had two-ply mist-nets designed for catching birds, and they were not suitable for catching insectivorous bats, but we caught Zenker's Fruit Bats and the tiny nectar- and pollen-eating Woermann's Long-tongued Fruit Bats, these being the only species of fruit bats that inhabit the lower strata and clearings within the rainforest.

The second trip was in July, to a cave near the remote village of Ipole, near Ilesha, roughly 100 kilometres east-north-east of Ibadan, where we had heard that fruit bats were roosting. In Africa, only three species of fruit bats are known to roost in caves and only one of these, the Egyptian Rousette, was known from Nigeria. However, it was known from only two areas and appeared to be rare in Nigeria although common elsewhere. Consequently, we were keen to check the bats in this cave. So we drove to Ipole accompanied by two zoologist friends. At Ipole, we found a local Nigerian who was willing to guide us to the cave. It was a five-kilometre walk along a narrow path that led through areas of rainforest, small fields of local crops, and some groves of cocoa trees. We met two Yoruba hunters on the path who were dressed in traditional garb and carried dangerous home-made guns of an obsolete design known as Dane guns. They were hunting for monkeys and bushbuck or anything else that could be sold as bushmeat. Then we climbed down into a cool, fern-clad gully with tall rainforest trees and a stream. There were several waterfalls, and the entrance to the cave inhabited by the fruit bats was partly concealed behind one of these falls. We set a mist-net across the entrance of the cave and then went in to explore. We found the bats roosting in two clusters in dimly lit parts of the cave about twenty metres from the entrance, and we estimated that there were thirty-five to forty individuals. We netted twenty-two of them and then they were weighed, measured and described, and their reproductive condition was assessed. We had not really expected to find Egyptian Rousettes at this locality and we didn't. Our bats were Angolan Collared Fruit Bats, and this was an exciting new record for Nigeria.

Upper Ogun Game Reserve

David and I also had a very memorable trip to Upper Ogun Game Reserve with Hilary Sydenham from the Zoology Department and some of his assistants. Hilary was a freshwater fish biologist and he had been studying the fish in the Ogun River for several years. During the day, he and his assistant set nets across the river, and encouraged me to catch anything I could with a rod and line. While I was fishing, David went for a walk along the river: he had often walked there alone but, to me, it seemed that he was not very concerned that there might have been lions or crocodiles! I had the worse scare of my life that afternoon. I heard the most awful screaming and I thought it was David being killed by a lion or dragged into the river by a croc! Thank goodness it was only an Olive Baboon and I doubt if it was hurt either. It was I who nearly died – of fright!

Not long after dark, Hilary went out in a small canoe to check the nets, and David and I went with him. Hilary said it was perfectly safe but all around us we could see the red eyes of dozens of baby crocodiles gleaming in the light of our torches! I wondered where their mothers were, but Hilary was not the least bit concerned. He caught a lot of fish, most of which he took back to U. I. for measuring and dissection, but we had fish for dinner that night! We chose some freshwater Elephant Fish of the family Mormyridae, which produce powerful electric shocks to stun their prey, so they have to be handled very carefully. Hilary's assistant cooked them in palm oil and served them with rice, and it was the best fish dinner I have ever had.

Time to leave

In many ways it was sad leaving Nigeria, but we were confident that our decision to leave was the right one. Soon we were boarding our plane and were on our way to Salzburg in Austria, our first destination in Europe before returning to Australia. We stared out through the windows as the Sudan savanna with its little villages gave way to the dry, sparsely populated Sahel savanna and then the bare sands and occasional rocky outcrops of the vast Sahara. Our feelings of sadness were only temporarily lightened by the Nigerian captain of the aeroplane who told us, over the intercom, that, "We 'ave a strong headwind dat is pushin' us backwards."

7

MORE POSSUMS, A DOG NAMED MOSHI AND MAGPIES IN CANBERRA

The headwind did not push us backwards after all, and we *did* get to Salzburg in glorious Austria, where we enjoyed a holiday with friends. Then we had almost six months in England, seeing friends and relatives, and preparing to sell our house, before moving to Canberra. That was my first experience of late autumn and early winter in England, and it was so beautiful! On the day we drove to London to stay with David's uncle before catching our flight to Australia, there was a hoar frost that coated every tree, every blade of grass, every fence and every building with frost that glistened in the pale sunlight. It was magical.

Arrival in Canberra

We had mixed feelings when we arrived in Canberra. For me, it did not feel like coming home because it was not Victoria. And for David, it was neither England nor somewhere in Africa. But Canberra had many advantages. It is a well-designed, new city with a country feel about it. It has a lovely central lake, many parks, houses with large gardens and lots of trees, wide streets with wide nature strips planted with trees (exotic and native), suburbs with local shops, schools and other facilities, several town centres with large shops, department stores, offices, theatres and so on, and government institutions where large numbers people are employed.

Canberra is also home to the Federal Parliament and most of the federal government departments.

We were housed in a flat until our furniture and other possessions arrived from Nigeria. Then we started hunting for somewhere else to live and we found a double-brick house with an extra-large garden in Spencer Street, Turner – and we bought it. It was within easy walking distance to the Australian National University.

In the seven years that followed, David began research projects on animals in the nearby mountains, the Haplets started pre-school and then primary school and I was able to do some part-time teaching (demonstrating) in the Zoology Department at the Australian National University. Our lives at No. 1 Spencer Street were enriched by friendships with many animals. These included some ducks, chooks (full-sized hens) and bantams, several Brushtail Possums, several Australian Magpies, and a very *very* special dog who was like another child to me.

Our first possums in Canberra

The daughter-in-law of our next-door neighbours was a vet and she asked us if we would like to look after a Brushtail Possum that had been attacked by a dog: the nerve running down the possum's arm to her left paw had been bitten and this limb was paralysed. The vet thought that the possum should be kept in captivity for six months because, during this time, either the nerve would regenerate or the possum would learn how to manage without a functional left arm. The nerve did not heal after six months but the possum did learn to cope fairly well without it – but then she escaped and we did not see her again.

During her captivity, we kept this possum in a large enclosure inside a conservatory into which the back door of the house opened. But the conservatory was not 'possum-proof' and some of the wild possums in our garden often came in to investigate the caged one. One was a female with a joey on her back. We offered her food and she became quite tame and her baby, Shadow number two, became *very* tame. This Shadow was ear-tagged and she remained in our area for eleven years. We had a lot of international visitors in those days and Shadow was a star attraction. She would always come if she were close enough to hear us calling her, and then she would sit on my lap and allow me to open her pouch so our visitors could see her pouch-young if any were present. She took food from our hands and even the visitors could stroke her.

The other visitor to our captive possum was an adult male who was very nervous of humans and he never became tame. However, one night, he turned up in the conservatory with a two-centimetre-long wound on one shoulder – we don't know what caused this injury. The wound looked healthy at first but, after a few days, the possum came into the conservatory again and we saw that the wound had become badly infected and the possum looked very sick. We had some antibiotic powder from our vet friend, and on this occasion the possum stayed still when I approached him with food in my hand, and I was able to spray the wound with the powder. Incredibly, the possum came back almost every night for about a week, and let me put powder on his wound, and then the wound healed and we did not see him very often after that. What made him keep coming back to have his wound treated?

Moshi

David and I felt that our family would not be complete without a dog, so we started debating the pros and cons of different breeds and then decided that we would like to have a black Labrador. We could not afford to buy a well-bred dog from a breeder, so we were very interested when some Labrador puppies were advertised in the local paper, and we drove to another Canberra suburb to look at them. We were introduced to the golden Labrador who had had the pups, and we were impressed by her temperament. We were told that she was about twelve years old and that a vet had concluded that she would never be able to have puppies because she had been injured when she was young. Consequently, she had not been spayed and she was allowed to wander at will when she was in season, and nothing had ever happened – until this time! We were told that the mother had been observed mating with a black Labrador who lived in the neighbourhood, and it was for this reason that the pups were advertised as Labradors. And indeed, to our inexperienced eyes, the eight-week-old puppies looked just like Labradors. We chose the one who played mischievously with the curtains and was happy to play with the Haplets. Although some of the puppies were golden like their mother, this one was black and had a smattering of white on her chest. She cost us ten dollars – the best ten dollars we ever spent.

Although she was only eight weeks old, we brought our little treasure home and she settled in immediately without appearing to miss her doggie family. We decided to call her Moshi, which is the Swahili word for smoke. We wanted to have an African name, and 'smoke' was appropriate because I had known a lineage of wonderful black Labradors whose names were Smoke, Cinders, Soot and Smut.

Moshi was the best dog I have ever known. She was highly intelligent and absolutely trustworthy, but her two most outstanding characteristics were that she always wanted to please us, and she thought she could do anything that humans could do. She would sit on the Haplets' tricycle and on their rocking-horse. She would allow herself to be clothed in jackets and trousers, and she would lie on her back in a pram while the children wheeled her around like a baby. She would sit with them while they played at being at school and, on one occasion, she even climbed a tree with a sloping trunk because that's what the children were doing! The discovery that she could not turn around to climb down again *did* teach her that dogs are not meant to climb trees!

But watching Moshi growing up was somewhat disconcerting. Her head, body and tail grew as one would expect the head, body and tail of a Labrador to grow, but her legs remained very, *very* short while her ears grew bigger and bigger and began to stand alertly erect! Eventually, she reminded us of the sort of animal one finds in children's books where the pages are cut into three sections so that the top, middle and bottom sections of different animals can be interchanged. Our little Moshi had the huge, pricked ears of a Corgi, the head, body and tail of a Labrador, and the short, very dexterous legs and paws of a Corgi. If the mother had had a romance with a black Labrador, evidently she had also had one with a Corgi!

Moshi had characteristics of both parents. She loved water and swimming and she became addicted to fetching things that were thrown for her entertainment. But she also had the herding and guarding instincts of Corgis that make them useful as cattle dogs, and also the manual dexterity that makes them such renowned rat-catchers.

Moshi's willingness to please us made her wonderfully easy to train. She loved to learn new things and, as soon as she learned to recognize when we were trying to teach her something new – by teaching her a new word associated with a particular behaviour – there was no limit to what she could be taught. We always combined commands with hand-signals, and she learned to respond to one, or the other, or to both. For example, like all young dogs, Moshi loved chasing her own tail and catching it. Whenever she did this in my presence, I said "Tail" and made a small, circular movement of my first finger. Before long, she associated the word and signal with the activity so that I could elicit the behaviour by saying "tail" and by circling my finger. We could entertain people by asking Moshi "Have you got a tail?" With no signal from me, this would elicit no response other than a questioning alertness. Then I would say, "Where is your tail? Catch it" and with these words I would give the signal and Moshi would spin around and come up with her tail between her teeth.

Moshi learned an impressive number of tricks and games as well as all the behaviours that made her a joy to have in our lives. She always came when called, no matter what she was doing at the time. She would stay still in one place until given the "Off" command, which gave her the freedom to do whatever she wanted to do, and she would beg, with her front paws tucked into her chest, for treats. Very soon, the beg posture signified that she wanted us to *do something* – not just to give her an edible treat. She loved to play games and a great favourite was 'Hide and Seek' – on being told, "Kitchen" she would take herself out of sight (either in the kitchen or somewhere else) while we hid a special piece of wood and then, when told, "Find it" she would search enthusiastically until she found it.

One game, however, became an obsession! Above all things, she loved having sticks thrown so that she could chase and fetch them. She would bring sticks to us – or anyone else – lay them at our feet and then beg to request us to throw them. For this obsession, anyone was fair game, and no one could walk past our house without being presented with a stick that they were meant to throw. Eventually, this obsession became annoying and so we decided not to throw any more sticks.

Moshi was marvellous with all our animals and birds. One day, Lena brought home a fledgling Australian Magpie whose parents had been killed on a road, and a very strong bond developed between this magpie and Moshi. The pair invented games that they would play every day. One was a sort of 'follow-the-leader'. It was only played inside the house and it involved one following the other around the legs of the furniture, under the tables, around the sofa and so on. Sometimes Moshi was the leader but, as often as not, the leader was the bird. Another game was typical of that played by puppies and young dogs – Moshi would adopt the well-known 'bottom-up' posture with the front legs splayed out on the ground, and then one of them would pounce at the other while the other jumped away. Sometimes it was Moshi who pounced at the magpie and sometimes the bird would pounce at the dog. There were two rules that applied to all of the games and interactions between the dog and the bird – Moshi never opened her mouth, and the bird never flew.

We played a game with Moshi that involved us sitting, with our legs wide apart, at opposite ends of a large rug. Moshi would stand in the middle, and we would try to roll a tennis ball past her and into the gap between our legs. Moshi, of course, would try to intercept it. The magpie was fascinated by this game and tried to join in, so we replaced the heavy tennis ball with a ping-pong ball and the bird soon became a keen player, with or without Moshi.

It was intriguing that, although the leadership role in the 'follow-the-leader' game alternated, the magpie was really the dominant one of the pair. We wonder if it thought of itself as a dog because it ousted Moshi from the basket she slept in, and it would lie down, on its side, in the basket with its legs stretched out, just like a dog! It would also roll onto its back to have its tummy tickled!

The friendship between Moshi and the magpie continued for many months, by which time the bird had learned to fly strongly and fend for itself in the neighbourhood. It often flew well out of sight of our house, but it would return and perch on the branch of one of our trees. As soon as Moshi spotted it, she would beg underneath the bird's perch, and then the bird would fly down and they would play together.

This incredible friendship came to a sad ending for Moshi. The magpie, which we had colour-banded so we would always be able to recognize it, left our neighbourhood and took up residence in a park some distance away. But the local Spencer Street magpies would fly into our garden and perch in the trees, and poor little Moshi could not tell that they were not her friend. Joyfully, she would run and beg to them, but, of course, *they* never came down to play and Moshi would creep away with her tail between her legs, looking thoroughly hurt and miserable.

In those days, we kept ducks and chickens, and the ducks were usually let out during the day to forage for snails and slugs in the garden. At dusk, we often made use of Moshi's herding instincts to round up the ducks and drive them back into the fox-proof chicken run. Lena also kept rabbits and occasionally let them loose in the garden where they often became lost in the vegetation. But Moshi could always find them, and she could quietly approach them and then immobilize them by resting her head across their necks until someone came to pick them up.

As well as her friendship with the baby magpie and her tolerance and reliability with our ducks, chooks, rabbits and various possums, Moshi developed a strong bond with a horse that I broke in. The horse – a mare named Paquita – was kept at 'Stonehenge' – a property on the outskirts of Canberra, where there were stables, yards and paddocks and also an area where landscape gardening supplies were for sale. A man who worked there had a young German Shepherd who loved playing with Moshi whenever I took her to Pialligo, but he was a bit too big and boisterous for Moshi and, one morning, she escaped from him by curling herself up between Paquita's front legs while the horse was lying down in one of the yards. As if this were not sufficiently amazing, the frustrated German Shepherd ran around and around the horse, barking and barking, but the horse dropped her head over Moshi and stayed put!

I only took Moshi with me on one occasion when riding Paquita outside 'Stonehenge' and that was memorable for one reason in particular. It had been very wet and there were some large puddles across the track we were following. Up to that time, Paquita had been very nervous of puddles and had refused to go into them. But Moshi loved water and plunged into the puddles with delight – and Paquita followed her and never minded going through water again.

The Spencer Street magpies

A pair of Australian Magpies inhabited the area around our house in Spencer Street and, every spring, they built a nest of sticks in one of the mature eucalyptus trees that had been planted along the nature strip in the 1940s. Australian Magpies are very different to Eurasian Magpies, which they resemble only in being black and white. The Australian Magpies feed on insects, spiders and grubs – never on baby birds – and their carolling is most melodious.

During our first springtime in Canberra, the Spencer Street pair had some problems. The male had been hit by a car and could not fly. The female solved one consequence of this by building a nest that the male could reach by climbing the sloping support wire of a lamp post and then walking along the electricity wires until he could jump off onto a branch and then walk along the branch to the nest. According to an old forester who lived in the street, it was the flightless male who sat on the nest and incubated the eggs, and the female who brought most of the food to the chicks. If this were so, it was abnormal behaviour because it is usually the female who builds the nests and incubates the eggs. With this strategy, the pair raised two fledglings – but then another problem arose. Both of the young had abnormally tight tendons in their feet and, for several weeks – until their tendons stretched – it was difficult for them to walk. This was serious because young magpies do not fly when they first leave the nest. They learn by copying their parents but, in this case, because the male could not fly, the young did not attempt to fly and so they were very vulnerable to attack by cats, especially at night when they should have been safely roosting in trees. Rightly or wrongly, I decided to intervene by catching the babies and bringing them inside during the night, and then putting them outside again in the morning – and I also threw them into the air as often as I could so they would learn to fly. However, catching the babies was hazardous because adult magpies can become quite aggressive during the breeding season and they often swoop down on people – usually from behind – and scare them by clacking their beaks just as they are swooping past. Sometimes they will actually peck people's heads too. Because of this, I always attempted to catch the babies when the adult female was somewhere else, but I was caught 'red-handed' one evening, with both babies in my hands. The female launched a very serious attack and pecked a long gash across my scalp, which, of course, bled profusely! But I managed to hold onto the babies and quickly got them, and myself, safely inside the house.

The next morning, I wore a hat when I took the babies into the garden to be released but, although the female was present, she did not attack. I had a handful of chopped meat with me and threw some to the female who immediately flew down to eat it. Within a few days, a routine was established. I would carry the babies outside and then sit on the lawn with my legs apart and the babies between my legs. I would hold meat in my right hand, and the female (or sometimes the male) would take the meat, hop over my leg and then feed the squawking babies. I very soon stopped wearing a hat because I was never swooped again – even if I were catching the babies and walking somewhere with them in my hands!

As their short tendons lengthened, the babies rapidly became very mobile on the ground, and I could expect to find them in any of the gardens along Spencer Street. It was a wonderful introduction to all of our neighbours and, at dusk, I could expect a phone call from someone to tell me that the baby magpies were in their garden. Throwing them into the air worked too, and it was not long before they learned to fly and then it was no longer necessary to bring them in at night. The male also recovered his ability to fly after he narrowly escaped being eaten by a dog! The whole family became our friends. They would fly down when we were gardening because we always found lawn-grubs where we were digging or weeding. Often, they were very nearly decapitated by our spades! The parental pair also discovered that they could get an early-morning feed by flying through the window into our bedroom, perching on the end of the bed and filling the room with the glorious sound of their carolling.

This Spencer Street pair was still around in mid-1984 when we went away to spend a year doing fieldwork in Malawi (Chapter 9). We rented the house to a man who assured us that he loved birds and would take great care of the magpies during our absence but, when we returned a year later, there were no magpies in our garden and we were very sad. However, a few weeks later, I was walking along Boldrewood Street, which runs past the south side of our garden, and I noticed a magpie fossicking in long grass beside the footpath and, without really thinking, I called "Maggie, Maggie, Maggle-pie" as I had always done to the Spencer Street magpies. To my amazement and delight, the bird emerged from the grass and came to heel like a well-trained dog! I knew it was our male because he had three short feathers that always grew awry on one shoulder as a result of him being hit by the car before we arrived in Spencer Street. Like a dog, he walked beside me all the way back to our house, and then waited by the back door while I fetched some grated cheese. He filled his beak with as much as he could carry, and then flew off across Boldrewood Street and into some trees about two hundred metres away. As he approached the trees, I heard the squawking of baby magpies and, as soon as they had fledged and become mobile, he brought them into our garden! We assume that these young had a different mother because they did not inherit the short tendons that had afflicted the male's earlier clutches and, of course, since both parents could fly, their babies learned to fly without my intervention.

The saga goes on. The poor old male was hit by another car several years later. Lena found him lying, unconscious, in the gutter with blood oozing from his nostrils. We thought it was the end, but we put him in a dark box in a warm room and waited to see what would happen. To our amazement, he regained consciousness and he was so tame that finding himself inside the house with us did not shock him, and he was ready to accept food by the following morning. However, once again, he could not fly! After a few days, we let him go in the garden. We knew our dog Moshi would not harm him, but we had another dog by then – our daughter's Chihuahua – and this one had to be kept on a lead whenever she went outside. Of course, inevitably, she managed to escape outside one day, and she was unable to resist chasing the magpie! It might have been the end of the old bird but, miraculously, he discovered that he *could* fly after all!

The old male survived for a year or so more, and then simply disappeared. However, his descendants are still around, and they still come carolling for food near the kitchen door and keep us company when we are gardening.

And no one near our house in Spencer Street is ever swooped!

8

'DIMBILIL' WITH MOSHI, A WOMBAT, NAUGHTY COCKATOOS AND A GHOST

In 1981, we bought forty acres (sixteen hectares) of land at the head of the Urila Valley in the country south of Canberra, and this was to give us enormous pleasure. The land was a hilly mixture of pasture and semi-cleared woodland, with forest along one boundary, and we named it 'Dimbilil' after a farm in the highlands of Kenya where patches of natural vegetation and farmland interdigitated, and both wildlife and domestic animals were accommodated. Near the top of a hill in the middle of the property, there was a workshop that we converted into a simple cabin with a ladder leading up to a mezzanine floor where there were beds for the Haplets. There were wonderful views in every direction from the cabin. To the east, we looked over a valley with a dam and then up to the top of a ridge. To the south, we looked through large old eucalyptus trees to the hills at the head of the Urila Valley. To the west, we looked across our alluvial flat to forested hills on the other side of the valley, and to the north we looked right down the Urila Valley where there were many small farms of various kinds.

What we loved most about 'Dimbilil' was its diversity – there were so many areas with different terrain and different vegetation. The flat, almost treeless paddock in the front had a creek running through it, a stand of manna gums at its southern end, some deeply eroded gullies, and a frost-hollow with snow gums and black sallees at its northern end. And east of this flat area, there were hillsides with some very rocky areas and various eucalypts, groves of dense acacias, and the dam in a valley where there were taller trees and lush green grass. So it was not surprising that 'Dimbilil' was inhabited by Eastern Grey Kangaroos (as many as 170 in the front paddock during a drought), Swamp Wallabies and Red-necked Wallabies, Common Wombats, Echidnas, Brushtail Possums, Sugar Gliders, Bush Rats, Brown Antechinuses and many bats including Common Bentwing Bats and Lesser Long-eared Bats. It was also a wonderful place for birds – we recorded at least sixty species, and these included many that we did not have in our garden in Canberra. We were especially pleased to have Superb Lyrebirds in the forest next to 'Dimbilil'. It was so enjoyable listening to their songs and trying to recognize all the birds that were being mimicked.

Building a balcony

The cabin was built on a slight slope, and it stood on brick pillars to prevent attack by termites. Consequently, five rickety steps led up to a small landing in front of the front door of the cabin, whereas just two steps led down from the back door. By 1992, the front steps and landing had become too rickety for safety, so we decided to replace them. We decided to make the landing larger – and, by the time we had finalized our plans, it had become a balcony about five metres long and three metres wide! Because there was no power at 'Dimbilil', I cut all the timber we needed for the balcony, and the steps leading up to it, at home in Canberra. We spent a lot of time carefully measuring what was needed, and I enjoyed cutting all the timber, cutting all the joints, and sandpapering everything – with particular attention to the railings around the edge. These were beautifully rounded and then all the wood was painted dark brown to match the doors and window frames of the cabin.

We set stirrups in concrete to house the upright posts, and then built the balcony. Miraculously, we had not made any mistakes with the measurements, and everything fitted perfectly. The decking timber was nailed into place and treated with preservative, and a flight of four steps was built to lead up to the deck. A final coat of dark brown paint was applied where needed, and we went home to Canberra feeling proud and delighted with our achievement!

One week later, we went out to 'Dimbilil' and found that a flock of Sulphur-crested Cockatoos had found the treated-pine balcony rails – and also the front windowsills and front door of the cabin – an irresistible temptation to peck to pieces with their very powerful beaks! We had to cap all the rails with metal, cover the hole in the door with metal, and hang wire-netting over every window!

Doodle, a very, very friendly wombat

We had a lovely neighbour living on a property adjacent to the north side of 'Dimbilil'. She was a Canadian lady who had joined the Wildlife Rescue Service, and she raised many Eastern Grey Kangaroos and Common Wombats whose mothers had been killed on the roads. At first, Val did not have any experience with baby wombats. They are delightful little animals and they become extremely attached to their foster parents. Val's first – a male she named Doodle – was no exception. He followed her everywhere, and he *loved* being picked up and cuddled. But, unfortunately, he got into the habit of giving Val a gentle nip on her legs as a way of asking to be picked up. This nipping was fine while Doodle was still a baby but, by the time he had grown too heavy to be picked up, the nips had become quite painful bites!

Doodle used to wander up to our cabin when we were there. He watched us building the balcony and, at that time, it was just possible for David and Jonathan to pick him up – but they said it was like lifting a twenty-kilogram sack of potatoes! Time passed, and Doodle grew. I particularly remember a hilarious evening when he came to the cabin to see us. All four of us went out to the back of the cabin to feed him some pieces of carrot. He enjoyed the carrot very much, but then he wanted to be picked up for a cuddle and that meant that one of us was about to be bitten. We made a dash for the back door and all of us, except Lena, managed to get inside before Doodle caught up with us. Undaunted, Lena set off to run, in a wide circle before coming back to the cabin. Doodle watched with evident alarm. "If she is bolting, there must be grave danger and I had better bolt after her!" We shouted encouragement to Lena, and she outran Doodle. We had the cabin door open, she leapt inside and we slammed the door shut. But then we felt sorry for the poor wombat, so I suggested we fed him some more carrot – from the safety of the balcony. We had assumed, of course, that wombats could not climb up steps – but Doodle could! And did! Lena and I hurried inside the cabin, but David and Jonathan – like old maids with a mouse in the room – leapt onto the table that we kept out on the balcony. Lena and I were convulsed with laughter! But Doodle had the last laugh. I went out of the back door and around to the balcony to entice him to come down. He came down, determined to make me pick him up for a cuddle. I beat a very hasty retreat, but I had forgotten that our car was in the way, and Doodle was able to corner me and he bit my leg very hard!

Intruders at night in the cabin

Occasionally, we were woken up during the night in the cabin by unexpected noises. There were sounds of small things being moved on the kitchen benches, or sometimes pushed onto the floor, and we often heard plastic bags or paper bags being shuffled about. The culprits were Brown Antechinuses – formerly known as marsupial mice although they are not rodents – which came into the cabin in search of food. These tiny, mouse-like animals with very long snouts are mainly insectivorous but they seemed to be attracted by bread, cheese and other scraps. We did not mind having antechinuses in the cabin because they ate pests such as cockroaches, silverfish and spiders.

On one other occasion, we were woken up by the sound of a ghost clanking chains as ghosts always do in the dead of night! Everyone knows that – but how many of us have actually heard this very recognisable sound?

I switched on the torch on the bedside table. Nothing to be seen although there were a few more clangs and clanks. David switched on his torch too, and then he lit the portagas lamp. Eventually, the ghost was spotted, flitting around in the shadows under the high-pitched roof of the cabin. It was a bat! We stayed very still and it started to fly closer to us, and then it actually landed on my pillow and I was able to catch it in my hands! Unfortunately, I cannot remember what kind of bat it was, but its communication calls – which are audible to humans – produced weird echoes in the large space of the cabin, and I am sure this must explain the belief that ghosts are often draped in clanking chains.

Moshi at 'Dimbilil'

Moshi adored coming to 'Dimbilil' and she would become very excited when she heard that word. At 'Dimbilil', she participated in almost everything we did. If the Haplets were transporting leaf-litter and sticks away from the vicinity of the cabin, Moshi followed them and then rode back in the wheelbarrow. She loved swimming in the dam and loved coming for walks with us and, in the early days before she became obsessed with fetching things, we spent a lot of time throwing sticks for her to retrieve.

In the early 1980s, there was a very severe drought and our neighbours with livestock were in deep trouble. At the time, because we had no livestock, we had more grass, so we volunteered to look after an old grey horse who was otherwise going to be put down. Her name was Cygnet, and she had been a great show horse in her prime. Moshi was very good with horses and she soon made friends with Cygnet – but, when we no longer threw things for Moshi to fetch, what did she do? She took a stick to Cygnet, dropped it at her front hoofs, and then begged! She looked straight into Cygnet's eyes, then looked at the stick, and then looked back into the horse's eyes – a dog's way of pointing to something. And once, when Cygnet failed to respond as usual, she picked up the stick and whacked it against her hoofs! That was too much for Cygnet who promptly galloped away. Moshi was disgusted!

Moshi was often a great help to us at 'Dimbilil'. On several occasions, I went down the hill from the cabin to the dam to dig up thistles but forgot to bring heavy leather gloves to guard against the vicious prickles. Of course, Moshi came with me and that was very handy because I could send her back to the cabin to "Find the Master". Then I could shout, "Gloves" to David, and he could give them to Moshi, say "Find the Missus" and she would race down the hill to bring them to me. Things like that were so useful. Moshi could take written notes to David or me – or anything else we wanted her to take. She was like another child to me and having her put to sleep in 1993 was terribly hard. But, by then, she was deaf, blind in one eye and needed a special geriatric diet and we were going away for year, so we knew it was time to say goodbye. It is given to us to spare our animals the suffering that so often comes with old age – and this is a gift we should surely exploit to give a pain-free ending to the lives of those who have given us so much love in their time.

Moshi is still at 'Dimbilil' and when we sold this wonderful place in 2009 (when weed control, another drought and the threat of bushfires were getting too much for us), it was going down to her grave for the last time that unleashed our tears.

9

FIELDWORK IN WILD PLACES, AMAZING BATS, AN ELEPHANT AND OTHER ANIMALS IN MALAWI

Introducing Malawi

In 1984, after his first seven years at the Australian National University, David was granted twelve months of sabbatical leave, but it had to be from mid-year to mid-year so that he could squeeze all of his annual teaching commitments into the six months before and after his time away. Because he had already worked extensively in an African desert (the Sahara in the Sudan) and in African tropical rainforest (in Nigeria), he was keen to work in African savanna and montane habitats, so Malawi was an obvious place to go to. We could be based at Chancellor College (the university in Zomba) and we could make almost year-long studies in a mosaic of montane grassland and montane forest, and in two distinctly different types of savanna. The Haplets had to come with us of course, and we assumed that formal education would be impossible if we had to keep moving from one study area to another, so we decided that Lena (aged eleven and in grade six) would come back to grade six and carry on where she left off, and that Jonathan (aged nine and in grade four) would simply miss a year.

Malawi is situated in central Africa and is surrounded by Tanzania in the north and north-east, Zambia in the north-west, and Mozambique in the south-east, south and south-west. It is a small country by African standards, only about nine hundred kilometres from north to south, and

80–160 kilometres from east to west. Its dominant feature is Lake Malawi (29,600 square kilometres), which flanks most of the eastern side of the country. This very deep lake lies in the Western Rift Valley and it is drained by the Shire River, which runs south through an extension of the Rift Valley known as the Shire Valley, until it joins the mighty Zambezi River. West of Lake Malawi, the land rises steeply and is part of the Central African Plateau. Within Malawi, the Central African Plateau is flattish except for some mountains high enough to support montane grasslands and montane forest. These include the Nyika Plateau (1800–2500 metres) in the north, the Viphya Plateaux (1500–1800 metres), Dedza Mountain (2198 metres) and the Kirk Range (1400–1600 metres). South of Lake Malawi, the Shire Valley is flanked by the Mangochi Highlands (1000–1500 metres), the Shire Highlands (900–1200 metres) dominated by Zomba Plateau (1500–2000 metres), the Phalombe Plain (700–900 metres) and Mount Mulanje (1000–3000 metres above sea level).

Malawi has a hot wet season from November to early April, a cool dry season from April to July and a hot dry season from August to October. The duration of each season varies with latitude, the Lower Shire Valley in the south having the shortest wet season and the longest hot dry season. The mean annual rainfall varies from 775 millimetres in the Lower Shire Valley to 2164 millimetres on and around Mount Mulanje, and temperature regimes range from extremely hot in the Rift Valley to cool on the high plateaux. The vegetation, although complex and varied, can be classified into ten main types of which the most prominent are the miombo woodlands (dominated by *Brachystegia* trees), the lowland lake and river plains woodlands, the mixed escarpment-foothill woodlands, and the montane grasslands and montane forests.

First impressions of the Warm Heart of Africa

We flew to Malawi from England where we had been visiting friends and relatives and working in the Natural History Museum where there are good collections of the mammals from Malawi that we needed to be able to recognize.

I was very nervous about arriving at Lilongwe Airport having experienced the chaos and corruption that had been so rife in Nigeria. We had so much luggage that was crucial to the success of our sabbatical – four heavy rucksacks (full of reference books and various pieces of equipment such as cameras, films, callipers, and special pens and pencils for making illustrations) that had come in the cabin with us, and four cases, four sausage-shaped bags (containing tents, sleeping-bags and clothes) and a bag containing other equipment, which we hoped had come in the hold. We had ninety-two kilograms of luggage altogether. We waited anxiously to see the hold luggage unloaded. Many other people got theirs fairly soon and started going through customs, but we watched and waited and nothing arrived for us! Then a lady from one of the customs desks approached me, looking very serious.

"I am very sorry about your luggage," she said solemnly.

"What do you mean?" I gasped.

"Well, it hasn't come off the plane, has it?"

"No," I replied, "but the plane is still on the tarmac so why can't someone check if it is still on board?"

"The plane is ready to go to Mozambique now – but perhaps your things can be dropped off on the return journey."

"But we need them *immediately*!" I said in despair.

She shrugged her shoulders and said, "It might even go all the way back to London!" But I saw her lips twitch and then she could no longer keep a straight face. I wanted to hug her! I suddenly realized how different the Malawians were to the Yorubas of Nigeria. They were my kind of people, with a great sense of fun, and I felt completely at home with them from then on.

Our luggage arrived in due course and the mischievous lady saw us through customs without raising any problems. Outside the airport, we were met by an expatriate lady who had come in a VW Kombi that we had bought, second-hand, from her because she was leaving Malawi. She drove us to the Lilongwe Hotel along a wide avenue lined with flowering trees, gaudy bougainvilleas and beds of cannas and other tropical flowers that we remembered from the gardens of expatriates in Nigeria. But there the similarity ended. Lilongwe was a new city, much influenced by the layout of Canberra and other planned cities, and it was scrupulously clean and tidy. We were much encouraged and, especially for David, it was wonderful to breathe the dry, scented African air and to hear the cooing of African doves – the most memorable sound of Africa.

We had lunch at the Lilongwe Hotel – our first taste of the '*Chambo* and Chips' for which Malawi was famous. *Chambo* are fish from Lake

Malawi and, in those days, fillets were about twenty centimetres long and absolutely delicious. It was also at the Lilongwe Hotel that we learned our first words in Chichewa – the most commonly spoken language in Malawi.

"*Morni. Muli bwanji?*" Hello. How are you?

"*Ndili bwino. Kaya inu?*" I am well. And you?

"*Ndili bwino. Zikomo.*" I am well. Thank you.

"*Zikomo.*"

The next day, David drove the Kombi back to the airport to collect several more cases of equipment that we had sent as freight – and yes, it was there waiting for us. Then he went to the Forestry Department hoping to obtain the permits we needed for catching animals, and they were *not* ready – not *everything* works in Malawi! Most of the following day was also spent doing administration chores such as getting the Kombi registered in our name and insured, but we had everything done by lunchtime and then we set off for Zomba, about 340 kilometres away.

We travelled along a tarmac road across the southern part of the Central African Plateau, with Mozambique on the right side of the road. Being the dry season, everything was brown except for rocky hills that looked blue in the distance, and the hazy pale blue sky. This part of Malawi is rural and comparatively densely populated, and we passed many clusters of mud-brick houses thatched with tall, thick-stemmed grasses that are harvested after the wet season. The ground around every house was swept clean, and there was no rubbish anywhere. In 1984, the civil war in Mozambique was still raging, and there were many temporary refugee settlements along the eastern side of the Lilongwe-Zomba road. The road was lined with refugees who were trying to sell kerosene lamps, bottles of cooking oil and other goods that international aid organizations had sent to Malawi. At intervals along this road, we should have had glimpses of Lake Malawi, but the air was hazy with the smoke from burning stubble and grass fires that are lit to promote the growth of fresh young shoots. There was a glorious red sunset over Mozambique and, soon after, we drove passed Dedza Mountain and then descended the steep escarpment of the plateau into the Upper Shire Valley. It was dark before we reached the bottom. At Liwonde, we crossed a bridge over the Shire River, and then drove across the valley until we reached the escarpment leading up to the Shire Highlands. As it was dark, we saw nothing of the countryside except for long red 'necklaces' on the hillsides that were the fronts of the slow-moving grass fires.

About five hours after leaving Lilongwe, we reached our destination – the home of Dan and Liz Goddard who were dear friends from our years in Nigeria. They were now lecturing at Chancellor College. The Goddards had arranged for us to spend a few days in a neighbour's house that was temporarily unoccupied, and this gave us time to buy or borrow various necessities, hire a cook-*cum*-housekeeper, and prepare to move into a house on 16th Avenue on Old Naisi Hill on the outskirts of Zomba. This was a newish brick house that had been occupied by the British High Commission while Zomba was still the administrative capital of Malawi. It had three bedrooms and a bathroom leading off a small hall, and a living-room, dining-room and kitchen, and it was furnished with beds, tables and chairs. Outside, there was a lovely but sadly neglected ornamental garden, quarters for servants, and a garage with a concrete-floored annex. It suited us perfectly, which was just as well because it was the only house in Zomba available for renting!

We employed a dear old man from the northern part of Malawi to cook for us and to do all the housework, some of the shopping and, sometimes, to keep an eye on the Haplets. From the very beginning, we respected this very kind man and we never dreamt of calling him anything but Mr Longwe.

During our first days in Zomba, the Goddards introduced us to many families – both Malawian and expatriate – and we were often invited into their homes. These people had a supply of cutlery, crockery, cooking utensils, recipe books, sheets, blankets and all sorts of things that visitors to Zomba could borrow. And everyone wanted to tell us about the animals they had in their gardens – mongooses that stole their eggs, squirrels, monkeys and bats! The Bone family had lots of bats in the roof of their house and they invited us to come and watch at dusk when streams of these bats dived from a crack under the eaves and then flew away. We also met 'Sir', the wonderful 'Mr Chips' headmaster of the Sir Harry Johnston International Primary School in Zomba, and we arranged for the Haplets to spend at least one week every month at this school, at least for the first six months of our stay in Malawi.

We were delighted with Zomba. It was a very picturesque town with many old colonial buildings that were relics of the days when Zomba was the administrative capital of Nyasaland during the colonial era and also when Malawi first became independent. Many of the residential homes were also built in the gracious colonial style – especially those built along a series of parallel roads that skirted the lowest slopes of Zomba Plateau that towered above the town. These homes had wonderful ornamental gardens with mature trees such as jacarandas, flame trees, frangipanis, oleanders and mangoes. The roads through the older parts of Zomba were lined with jacarandas that were in full bloom when we arrived. There were also some stands of tall, white-barked eucalypts but, unlike those in Australia, their leaves were completely undamaged by insects. Zomba still had a much-used gymkhana club, a governor's residence where the president stayed during his visits to Zomba, an old cemetery where many famous early residents were buried, a monument to the King's African Rifles and many other very interesting old buildings. There were also many small shops, run mostly by Indians, where one could buy just about anything. Most of these shops had a wide verandah (called *khondi* in Chichewa) and many were used by *khondi* tailors who sewed everything from sheets to suits and dresses on old, treadle-operated Singer sewing machines that were independent of the somewhat unreliable electricity supply. Zomba also had two small supermarkets, the Wonder Bakery that baked bread and a few cakes, and a typically African market where one could buy meat, fish from Lake Malawi or Lake Chilwa, fruit and vegetables (including 'Irish' potatoes, carrots, strawberries and other non-tropical things that could be grown on Zomba Plateau), metal buckets, pots, pangas (machetes), grass-cutters and watering cans, woven baskets, mats and hats, enamel ware from China, and cheap plastic odds and ends.

Plans for our research in Malawi

During our first week in Zomba, we finalized David's plans for comparing the ecology, reproductive biology and demography of small mammals – rodents especially – in three habitats that had roughly the same latitude but very different altitudes and consequently very different climates and vegetation. The localities he selected were Zomba Plateau at 1900 metres, Liwonde National Park at 500 metres and Lengwe National Park at 100 metres above sea level. Our original plan was to spend almost one week per month trapping, marking and releasing animals at each of these localities, and to spend the fourth week doing something different such as visiting the museum in Blantyre or carrying out small mammal surveys at other localities. However, we had not anticipated a major shortage of petrol! It was so severe that it seemed probable that we would be stuck in Zomba without even enough petrol to get ourselves to the top of Zomba Plateau.

Fortunately, we had decided to bring mist-nets for catching bats, a key to the species of African bats, and some notes on the bats likely to occur in Malawi that I had made at the Natural History Museum in London. Our original idea was that I should catch bats at each of our study areas when the small mammal projects were finished for the day and, in this way, we hoped to obtain new distribution records for the bats of Malawi and new records for some of the national parks. The petrol shortage forced us to make contingency plans – we would start three bat projects in Zomba. The first was to survey the bats that occurred in the large, ornamental garden around our rented house. The second was to find out as much as possible about Banana Pipistrelles (*Pipistrellus nanus*), which roost during the day in the furled new leaves of banana plants, and the third was to study the reproductive biology of the Little Free-tailed Bats (*Tadarida pumila*) that roosted under the corrugated-iron roof of the Bone family's old colonial house in Zomba. However, as luck would have it, we always managed to get *just* enough petrol to keep up the monthly visits to David's study areas! People kindly lent us their spare jerrycans so we could store petrol whenever we managed to find an open petrol station. Sometimes we had to wait many hours in a queue before being served, and once, when we heard that a delivery was expected, David spent twenty-four hours in a queue! And David decided to camp on Zomba Plateau rather than make separate trips every day during his monthly checks. Lions did not wander over the plateau very often and we were assured that the leopards had easier prey to hunt than humans!

More people use the scientific names of African rodents, shrews and bats than their cumbersome vernacular names and, furthermore, the vernacular names for many species vary from place to place which can be very confusing. Therefore, scientific names are used from here on for the animals we studied. The scientific names of animals referred to by their vernacular names are given in Appendix A, and the vernacular names of study animals which are referred to by their scientific names are given in Appendix B. Similarly, trees and other plants mentioned in descriptions of special habitats, are referred to by their scientific names.

Zomba Plateau

On 4 September, we had our first excursion to Zomba Plateau. This tabular massif has steep sides and a basin-like plateau, about 1500–1900 metres above sea level, that was covered by natural montane grassland with montane forests in folds, valleys and along streams. There were also some extensive plantations of exotic patula pines.

On our way up, we passed a sawmill and stopped there to order wooden stakes for marking our trap-sites, poles for supporting mist-nets, and timber for making a collapsible table for fieldwork. Just past the sawmill, the plateau road divides into a tarmac 'up-road' and a gravel 'down-road'. Neither was wide enough for vehicles to pass each other, and the terrain sloped precipitously at the edges of both roads. The up-road wound through plantations of patula pines and then, as the terrain became steeper, through natural evergreen forest. As we gained height, we had glimpses down to Zomba and over its surrounding hills, villages and farmlands. There was too much smoke haze for us to see Lake Chilwa and Mount Mulanje although we knew they were there in the distance.

At the top of the up-road, we passed the Kuchawe Inn, which is a very popular tourist destination, and then we followed a narrow, dirt, forestry track that led up to Chingwe's Hole – a very deep vertical hole of unknown origin at the very edge of the plateau. The view from Chingwe's Hole is magnificent – one can look almost straight down into the Shire Valley with the Shire River winding through it, and then across to the Kirk Range on the other side of the Rift Valley.

There is a forestry reserve around Chingwe's Hole where a mosaic of montane forest and montane grassland is, at least to some extent, protected. It looked a promising area for David's high-altitude study and, two days later, we set ten exploratory trap-lines, each with ten live-traps. We got up at 0530 hrs the next day, and started checking the traps at 0700 hrs. We were very pleased to catch eighteen rodents representing four species – *Mus triton* and *Dendromys nyikae* in the grassland, *Lophuromys flavopunctatus* in bracken growing at the edge of the forest, and *Praomys delectorum* in the forest. A few days later, with the help of two forestry workmen, we set out two grids, each with forty-eight or fifty trap-sites that we marked with labelled wooden stakes.

Trapping on the plateau grids

David set live-traps on the two grids on Zomba Plateau for three consecutive days every month, and he marked and then released all the trapped animals. He also carried out monthly checks on the height, density and biomass of the vegetation (see below) and recorded rainfall and temperatures. He caught eight species of rodents and one species of shrew and they all had different habitat preferences. Of the three commonest rodents, *Praomys delectorum* was only found in the forest or adjacent tangles, *Lophuromys flavopunctatus* preferred unburnt grassland and bracken, and *Mus triton* preferred burnt grassland and never went into the forest. Four other species, *Dendromys nyikae*, *Dasymys incomptus*, *Otomys angoniensis* and *Pelomys fallax*, were also only found in the grassland. Population numbers fluctuated from nine to forty individuals on one grid, and from nineteen to thirty-one on the other. They were lowest in the late dry season, and then gradually increased during the wet season until they reached a peak in the early dry season. Each species had different survival rates and few individuals survived on a grid for more than five to six months.

Liwonde National Park

Next, we made a reconnaissance trip to Liwonde National Park, which promised to be the best place to study small mammals at the intermediate altitude of 500 metres above sea level. This national park conserves habitats typical of the Upper Shire Valley. It is flat, alluvial country with isolated rocky hills, and its habitats include reed swamps along the Shire River, floodplains with savanna grassland, mopane woodland dominated by deciduous mopane trees (*Colophospermum mopane*), mixed deciduous miombo woodland dominated by *Brachystegia*, *Sterculia* and *Kirkia*, and tall-grass tree savanna dominated by trees including *Combretum* and *Terminalia* with an understory of very tall grasses.

To reach Liwonde National Park, we drove north across the Shire Highlands and then dropped down over the escarpment into the warm, flat Upper Shire Valley. Everywhere was either black because it had just been burnt, or brown with the dead leaves of deciduous trees and tall, dried grasses. The white trunks of scattered *Sterculia* trees stood out in stark contrast. Just before we reached the Shire River at Liwonde township, we turned right along a laterite road that eventually crossed the railway to Beira in Mozambique, and then we entered the national park. We drove through miombo woodland, crossed a wooden bridge over the Likwenu River (which was just a series of pools at the time) and then reached the Administration Camp where there were offices, workshops and a village of small red-brick houses where the game scouts lived with their families. We introduced ourselves to some of the officials and met some of the game scouts who patrol the park. They were all very friendly and helpful, and suggested that our study areas should be located near the Administration Camp where there was less likelihood of us being disturbed by tetchy elephants. Most of the Liwonde elephants had migrated into the park from Mozambique where they had been traumatized by landmines and gunfire during the civil war. Consequently, it was not surprising that they had become very wary of humans and, in fact, a rash priest who had recently got out of his car to take photographs of some elephants, had been killed by them. We were also warned about lions because they were becoming more numerous. We had a drive around Chinguni Hill (that later became locally known as Happold's Hill) with one of the game scouts. We passed through tall-grass savanna, across over-grazed flats and through miombo woodland, and we were excited to see many Waterbuck, a small herd of Southern Reedbuck, a pair of Bushbuck, a male Lesser Kudu with incredibly long spiral horns, an old male Sable Antelope with long curved horns, and several Warthogs trotting along with their tails held straight up. This really seemed like the Africa we had read about and seen in movies! We picked out some likely areas for trapping small mammals, and then returned to Zomba feeling very optimistic.

Five days later, we returned to Liwonde National Park with Francis, a young man from Zomba whom we employed to help with fieldwork and to look after the Haplets when we were busy. Although there was an official camping ground where people were known to camp, we decided to pitch our tents near the offices in the Administration Camp because we thought lions and elephants were not likely to trouble us there. However, we had only just pitched the tents when the game warden approached to inform us that he could not permit us to camp there, or anywhere else, because there was a wounded lioness in the vicinity and she had been attempting to break into cars in her desperate search for food. We asked if we could stay overnight in one of the offices, but this was not allowed. Then, seeing that we were really desperate, the warden very diffidently offered us the use of an empty house in the game scouts' village. He never thought that we would accept, but we were delighted, not only because it was a safe refuge from lions but also because it gave us the opportunity to live in a purely Malawian village community! It was especially wonderful for the Haplets who learned more from that experience than they could ever have learned during the year they missed at school in Canberra. They learned how to communicate with children who did not speak much English, and they picked up quite a lot of Chichewa. Lena was considered old enough to take care of a baby, so she had little Elias strapped to her back with a wrapper – as did all the other young girls. Jonathan played football with the boys and learned how to make footballs out of plastic bags stuffed with grass and bound up with thin strips of bark from a particular kind of tree. He also went fishing in the Likwenu River with groups of boys and, in return, he entertained them by playing his recorder.

The house we were loaned was a simple two-roomed brick house with a small kitchen with a fireplace for cooking over, and a concrete workbench. Francis cooked his maize-meal over a fire, but we usually cooked everything on our tiny Trangia camp cooker, which had two saucepans, and we usually brought with us a cooked chicken and a frozen curry from Zomba as well as tins of meat that we could serve with rice or pasta. The house had a corrugated-iron roof, mosquito-screened windows and a concrete floor. A standpipe some distance away supplied water that had to be boiled before it was safe to drink, and there was a small outbuilding in which we could have a shower by pouring a bucketful of cold water over ourselves. There was no electricity, but we had our bushlights and torches for light, and we did not have a computer in those days. There was also no furniture in the house, but we had our collapsible table to work on and our foam rubber mats to sleep on, and we could sit on the seats that were part of the frames of our rucksacks. The foam rubber mats were comfortable enough, but I got stung by a scorpion that wandered over the floor on our second night, and that was disconcerting to put it very mildly!

After setting some exploratory trap-lines, David chose two sites for grids. One was a sloping rocky area with trees and short grass at the base of 'Happold's Hill' near the Administration Camp and the other was in woodland with very tall grass about 500 metres along the road that circles the hill. This area was also at the base of the hill and was also very rocky. Two game scouts (carrying .303 rifles made in 1917 during the reign of George V), came with us and helped to set up the grids and, from then on, we always had at least one armed scout with us when were working in Liwonde because of the elephants and lions. Luckily, we were never troubled by elephants, but we often heard lions roaring and, on one occasion when David was checking traps, he found the remains of a large male Lesser Kudu that had been killed on the grid during the night.

As the year unfolded, we caught seven species of rodents that were completely different to those on Zomba Plateau, and we also caught a very beautiful Dusky Sengi. Sengis (formerly known as elephant-shrews) are rat-sized insectivorous animals with delicate long legs and a very long, almost trunk-like, snout. They are actually more closely related to elephants than to either rats or shrews. All of the rodents were marked and released and, because they were often caught many times, we built up a picture of their habitat preferences, reproductive biology and survival rates. We also followed annual changes in the vegetation. We recorded grass biomass by cutting, drying and then weighing all the grass growing on five 50 × 50 centimetre quadrants. We measured the moisture content of the grass by subtracting its dry weight from its weight when first cut, and we measured the moisture content of the soil as well. We also measured the mean maximum height of the grass and, using a vertical 2.5-metre pole marked at ten-centimetre intervals, we devised an index of plant cover by counting how many of the ten-centimetre intervals were touched by the vegetation. These were simple procedures that we carried out at all of our three study areas, and they enabled us to describe important differences between each of our grids, and between each of our study areas. But it was often extremely arduous work, especially when the grass was long and tough, when the temperatures were high, and when clouds of tsetse flies were hell-bent on having us for dinner! We perspired so much at Liwonde that we never wanted to urinate during the day no matter how much we drank.

At Liwonde National Park, it was wonderful to experience the dramatic seasonal changes. It was the hot dry season when we started working there. The trees, except for a few palms on the claypans, were leafless and all the grasses were brown and dry. Some areas were just bare and dusty. Then the rains came in early November and everything changed. The miombo trees sprouted new leaves that were pink, orange or red, and the mopani trees were suddenly clothed in pale yellow-green leaves. Areas of bare ground became green parklands as grass began to shoot, and there were many pools – some clear, some very soon covered by floating weeds. Suddenly, the air was full of multicoloured butterflies of many kinds, dragonflies and damselflies, and the ponds were alive with tadpoles, caddisflies, fairy shrimps and water beetles. The heavy rain soon flattened the old tall grasses, but new shoots sprang up with astonishing rapidity. Liwonde National Park was transformed from an arid brown wilderness to a stunningly beautiful, green parkland.

But all too soon, the miombo leaves turned green, and the new grasses grew so high that they concealed most of the roads in the park – and most the roads also became impossibly boggy. Consequently, the park was closed to tourists, and we were glad that our grids were close to the Administration Camp. It became very difficult to see the marker-stakes on the grids and, at the beginning of each visit to the grids, David and his game scouts had to use pangas to cut pathways from stake to stake. The rain and the stifling heat and humidity made checking the traps extremely tiring, but the results were worth the effort. Then, during April, the wet season ended and from May to June there was no rain at all. This was the cool-dry season. The leaves began to fall (without turning colourful as in temperate autumns), and the grass browned off. The roads dried out and became passable again, and tourists were permitted to return.

Our study revealed that seventy-eight per cent of the trapped small mammals represented just three species of rodents – *Aethomys chrysophilus*, *Acomys spinosissimus* and *Gerbilliscus leucogaster*. Average population numbers fluctuated from three to seventeen individuals per hectare. The grid with the densest vegetation supported the greatest number of individuals. The highest population numbers occurred in the middle of the dry season (when the grids seemed to act as refugia) and in the early wet season (when young animals were old enough to move around the grids). Areas that had been burnt during the dry season did not support as many rodents as unburnt areas. Population survival rates were low, with only about half of the animals surviving on a grid to the following month. The number of animals per hectare was lower in Liwonde than in similar habitats in other parts of eastern Africa, probably because the wet season is shorter and therefore there are greater seasonal fluctuations in the availability of food, reduced opportunity for reproduction, and overall low productivity.

Lengwe National Park

On 25 September, we set off on our first trip to Lengwe National Park, the third of David's main study areas. Lengwe National Park, at 100 metres above sea level, conserves flat, alluvial country at the edge of the Shire River flood plain in the Lower Shire Valley. The climate is semi-arid and there are six main types of vegetation – forest, thicket, thicket-clump savanna, woodland savanna, tree savanna and grassland savanna. To reach Lengwe, we drove more or less south across the Shire Highlands to Blantyre and then down the escarpment into the flat Lower Shire Valley. The road down the escarpment is superbly engineered and has magnificent views. But, inevitably, it opened up country that had supported woodland and we were appalled to see hundreds of bags of charcoal for sale by the roadside and a landscape that was already denuded and eroding badly.

The Lower Shire Valley looked very arid compared to the Upper Shire Valley. We passed scattered villages with mud-brick houses with disintegrating grass-thatched roofs. There were goats and chickens wandering around, and a few scrawny cows. We often saw children and their mothers pumping water from wells surrounded by grey mud, and we often passed little tin or mud-brick sheds that had once been shops run mostly by Indians, but they were abandoned when the law changed and Indian trading became restricted to the big towns. We followed the tarmac road to the Shire River, which we crossed on a modern steel bridge, and then went straight on for another thirty kilometres until we came to the Sucoma Sugar Estate where there were vast fields of cane at all stages of growth. Here, we turned right off the tarmac road and drove along a dirt road that eventually led into the entrance of Lengwe National Park. In the dry season, this road was so dusty that the sugar company sprayed it with very sticky molasses! In the wet season, however, it became impassable and we had to take a much longer track across sandy ground to reach the park.

At the entrance to the park there were closed gates guarded by uniformed game scouts who checked our permits to enter the park without having to pay the usual visitor fees. When they opened the gates, these guards always saluted us with military panache – but, when they got to know us, they grinned mischievously from ear to ear at the same time. They were such fun!

Lengwe National Park had four chalets where visitors could stay. Each of them had three or four mosquito-screened bedrooms with beds and mosquito nets, electric lights, electric ceiling fans, chairs, tables and even a refrigerator! Sheer luxury compared with Liwonde National Park! There was a toilet-block near the chalets and we soon became aware of the all-pervading musky odour of free-tailed bats that were roosting, in great numbers, under the corrugated-iron roof. Most visitors disliked this smell but, being batty people, we were delighted!

Around the chalets, the ground had been cleared and swept, but there were some beautiful trees nearby, including some ancient baobabs. There was thicket-clump savanna not far away, and a path led through it to an artificial waterhole with a hide. In the dry season, it was the only water in the area and many animals and birds came there to drink. Lengwe National Park was primarily established to protect a rather special antelope – the beautiful, graceful Nyala – and we often saw herds of them at the water hole. We also saw herds of buffaloes plunging into the water, urinating and defecating in it, and turning it into a muddy mess, but the other animals did not seem to mind – especially not the Warthogs and Bushpigs who loved to wallow in the mud as all pigs do. We also saw Bushbuck, Common Duikers, Lesser Kudu, Yellow Baboons, Vervet Monkeys, troops of Banded Mongooses, and even a tiny Suni, one of the smallest antelopes. We often wondered why it was considered safe to walk, unaccompanied, from the chalets to the hide when there were so many buffaloes around, but at least there were no lions or elephants at Lengwe!

It was not easy to choose two sites for grids. The ground was almost devoid of grass and herbs. It had been overgrazed by the large herbivorous mammals, and millions of large termites (*Macrotermes* sp.) had cut tons and tons of grass and dragged it deep into their underground termitaria. We watched them doing this and marvelled at their industry! After some exploratory trapping, which was rather demoralising, we set up one grid in thicket-clump savanna with the same dimensions and number of traps as at Zomba Plateau and Liwonde National Park, and we set five lines of ten traps in five different habitats. It seemed silly to mark the grid with two-metre-high wooden stakes when most of the ground was bare, but the stakes made it easier to see the trap-sites that were in the thickets. However, the conspicuous stakes were strongly criticized by the resident biologist from Kasungu National Park on the Central Plateau who was visiting Lengwe at the time, but the Lengwe officials wisely allowed us to keep them and, later, we were extremely grateful that they did!

We were disappointed next morning, but not very surprised, when we only caught one mouse on the grid, but it was a beautiful little Griselda's Grass Mouse that was chestnut-brown with a dark mid-dorsal stripe. During the next few days, we caught a few more rodents on the grid and on some of the trap-lines, but we sincerely hoped there would be more when the rains came.

As at Zomba Plateau and Liwonde, we measured grass biomass, height and density every month. All of these parameters decreased very rapidly between September and December, but then the wet season began and, as at Liwonde, the trees and thickets burst into leaf and the grass began to grow – and grow – *and grow*! Before long, the tall marker-stakes that had been so conspicuous were completely hidden and, as we groped our way from one trap-site to the next, we often thought about the man who had wanted us to use short stakes. To make matters worse, the termites started to eat the stakes and we had to keep replacing them. During the wet season, we also had to contend with shallow floodwaters that made it impossible to set traps along some of the lines and, during one month, we had to wade through floods to reach the grid which, however, was never flooded.

One night at the beginning of the wet season, the large *Macrotermes* termites emerged in vast numbers for their nuptial flight, and this provided a food bonanza for bats, for a Greater Galago and also for the game scouts and their families. The winged termites were attracted to the lights outside our chalet and, with a butterfly net, we soon collected a bucketful that we were able to freeze. From then on, we could keep insectivorous bats in captivity and that proved very rewarding.

Because there were no lions or elephants at Lengwe National Park, we were almost always allowed to work in the field without having an armed guard with us, but we kept a very sharp lookout for buffaloes and we stayed in the Kombi if we saw any and waited until they had moved away. However, there was one time when Lengwe had another game warden, and he insisted that we took a guard with us. And this was the only time we came face to face with a buffalo! The guard was wise and calm. He did not want to shoot, so he made us back slowly behind a tree and then he scared the buffalo away by blowing a loud whistle that he carried for that purpose.

Altogether we caught nine species of rodents at Lengwe National Park. The commonest was *Mastomys natalensis*, a very feisty biting mouse that has twenty-four nipples and usually has ten to twelve young in each litter. Two other species were also common – *Acomys spinosissimus* (as at Liwonde) and a delightful grey mouse with a white belly called *Saccostomus campestris*. Less commonly caught were *Aethomys chrysophilus*, *Gerbilliscus leucogaster*, *Grammomys dolichurus*, *Grammomys ibeanus*, *Rattus rattus* and *Steatomys pratensis*, the Fat Mouse, which becomes very fat in the wet season and then aestivates in the dry season. Population numbers fluctuated from one to thirty individuals on the grid, and from four to fifty-seven individuals in total on the five lines. As at Liwonde National Park, there were two peaks in population numbers: the first in the early wet season when animals emerged from aestivation in dry season refugia, and the second in the mid-dry season when young born at the end of the wet season entered the trappable population. Survival rates were low with only thirty to fifty per cent of individuals surviving on the grid to the next month. The magnitude of the population fluctuations (especially for *Mastomys natalensis*) was related to the strong seasonality of the climate, flooding and grass characteristics and, in this, Lengwe National Park was similar to areas of Zambia and Botswana.

The demographic studies on rodents at Zomba Plateau, Liwonde National Park and Lengwe National Park were very interesting and, despite the problems with petrol, we were able to complete them. But, having started the projects on bats, it seemed sensible to keep them running.

Bat studies at Zomba

Our ornamental garden on 16th Avenue, Old Naisi Hill, on the outskirts of Zomba, proved to be a great place to catch bats partly because it had flyways through an avenue of old cypresses and around other large trees, and partly because there were other ornamental gardens, farmlands, and rocky hills with caves under piles of large boulders, less than one to two kilometres away. I decided to set mist-nets in the garden while David was camping on the plateau and, in total, twenty-one species of bats were recorded. They represented six of the ten families of bats known to occur in Malawi. There were two species of fruit bats (family Pteropodidae), which eat soft fruits, flowers and nectar. Fruit bats have very large eyes and find their way during the night by sight, and they are sometimes called megabats. All other bats in Africa are mainly or entirely insectivorous and use echolocation to 'see' their surroundings, and they are often called microbats. In our garden in Zomba, or nearby, there were five species of horseshoe bats (family Rhinolophidae), which have prominent structures (called noseleafs) on their faces to focus their ultrasonic echolocation calls into a beam in exactly the same way that a torch focuses light into a beam. Part of the noseleaf of a horseshoe bat is a parabolic reflector shaped something like the sole of a horse's hoof. There were two species of leaf-nosed bats (family Hipposideridae), which also have noseleafs but they are differently shaped, and two species of slit-faced bats (family Nycteridae), which emit their echolocation calls through a long slit running lengthwise down their faces. Slit-faced bats have enormous ears! There were also two species of free-tailed bats (family Molossidae), whose tail-membrane can slide down to the tip of the tail while the bat is flying, or slide up the tail so the tail-membrane can be furled close to the body. The wings of free-tailed bats are very long and narrow, which enables them to fly long distances in open spaces, but the wing-membranes can also be furled close to the body and, by furling the wing- and tail-membranes, these bats can use their forearms and legs for scuttling very rapidly over the ground or up and down vertical surfaces. Finally, there were eight species of vesper bats (family Vespertilionidae). Elsewhere in Malawi, four other families have been recorded – trident bats (family Rhinonycteridae), sheath-tailed bats (family Emballonuridae), long-fingered bats (family Miniopteridae), which we captured in other areas and one species of false vampire bat (family Megadermatidae), which has only been recorded from the extreme north of Malawi. Incidentally, there are no true vampire bats in Malawi and, in fact, no species of African bat feeds on blood.

Every bat that was caught in the garden, or anywhere else, was given a field number, and then its weight, forearm length, sex and reproductive condition was recorded in my field notebook. The vast majority of the bats were then released, but we had to keep some as voucher specimens and it was a condition of our permit to catch bats in Malawi that we lodged some specimens with the National Parks and Wildlife Department and the Museum of Malawi, Blantyre. Every specimen was given an accession number and then measured in more detail – head, total length, tail, ear, noseleaf (if present), tibia and hindfoot. An outline of a wing was traced, ectoparasites (living on the outside of the bats) were collected, and I also removed and preserved the kidneys because I was interested in renal form and function. By the end of our time in Malawi, we had recorded 1568 captures of bats although this included released individuals that might have been caught more than once.

In our garden in Zomba, the species most often caught was *Rhinolophus blasii*, which is often called the Peak-saddle Horseshoe Bat because the central part of its noseleaf rises to a high peak resembling the pommel of a peak-saddle used in western horse-riding in America. In October, I had eighteen captures of these bats so I started marking them so that recaptures could be recognized. Eventually, I had one hundred and twenty females and ninety-five males marked, and there were many recaptures including some marked in October that were recaptured in the following June. One very interesting result was that two females marked in January were netted again near a reservoir on the top of Zomba Plateau having moved from an altitude of about 900 metres to about 1800 metres. And later, one of them was caught again in our garden.

Twinkle, a Little Epauletted Fruit Bat

In Zomba, in early October, we caught a female Little Epauletted Fruit Bat (*Epomophorus labiatus*) who we kept under observation for a few days. We hung a branch of a mango tree in the hall of the house for her to roost in, and we fed her on various soft fruits, including bananas that she would not be able to feed on in the wild because of their thick skins. We called her Twinkle – "Twinkle, twinkle little bat! How I wonder what you're at!"– and she never seemed to be afraid of us. We wanted to photograph her but we had not yet unpacked our flash, so I suggested that we took her outside to photograph her in daylight. We thought that if I held her securely while David focused the camera, and then let her go, he might have time to take a photo before she took to the air. But no! She stayed clinging to my finger while David took all the shots he wanted. Then I decided to give her some banana so that we could photograph her with her cheek pouches bulging with food, and then, and only then, did she fly away. But where did she fly?

Round and round the house until she found an open window, and then straight back to her mango branch in the hall! We were astounded!

Ruff, Puff and other Peak-saddle Horseshoe Bats

The marking of the Peak-saddle Horseshoe Bats, *Rhinolophus blasii*, also enabled us to experience something even more astounding. In late December, we took two marked individuals into captivity in a room in the house and, for ten days, we recorded how many winged termites they ate, and how often they drank water when it was offered twice each day. These bats, Ruff and Puff, were able to fly around the room, but they were easy to approach and they soon learned to recognize the sounds of frozen termites being rubbed on the palm of my hand to warm them, and they learned to lap drops of water off my fingertip. After that ten-day period, water and termites were placed on the floor, and Ruff and Puff learned to fly down and feed themselves – so we kept them for another week before letting them go in the garden. Forty-five days later, I set four mist-nets in the garden, including one across the avenue of cypresses. 'Sir' from the children's school came for dinner that night, so I processed my captured bats as quickly as possible so I could spend about five minutes with 'Sir' before checking the nets again. This meant that, although I realized that I had a recaptured *Rhinolophus blasii*, I did not check which one it was until after I had let it go. Imagine how cross I was when I realized that it was Ruff and I had treated him like a stranger! I did not expect to catch him again because once a bat becomes aware that there is a net in any particular place, it emits its echolocation calls at an increased rate so that it can 'see' the net in plenty of time to swerve around, or over, it. Bats flying into nets are rather like parents falling over toys left on the floor: if they are looking carefully where they are going, they see the toys and avoid them. But the unbelievable happened! The next time I checked the net in the avenue, there was Ruff clinging to the net exactly where he had been caught about ten minutes earlier. And, this time, he had avoided falling into a fold and getting caught and therefore he could have flown away any time he liked. Instead, he let me pick him up and carry him back to my desk in the house. I put him down on the desk and started rubbing the palm of my hand to let him know that termites were coming. Then I fetched termites from the freezer in the kitchen and warmed them up

Mutual attraction and empathy – my brother and I with a foal when I was eight years old. (Photo: Gordon Stanley)

Hebe, my very best Christmas present. She and her descendants taught me how to establish rapport with animals. (Photo: Gordon Stanley)

Camidex, my adored thoroughbred horse.

A Fawn Hopping-mouse - one of the hopping-mice I dreamed about during my childhood.

A sand dune and claypan in the desert habitat of Hopping-mice on Sandringham Station in SW Queensland.

Collecting tadpoles, shield shrimps and other creatures in an ephemeral pool on Sandringham Station, with gibber plains in the distance. (Photo: Patricia Woolley)

Radio-tracking Heath Rats in their bushland habitat at Pomonal in Victoria.

Adorable Little Bandi - a free-living Southern Brown Bandicoot who became trap-addicted and then very friendly.

A Desert Mouse on a runway that it has swept clear of obstacles and bitten-off pieces of sedge.

School children
at Aroro, Nigeria,
with the live-traps
they constructed to
catch mice.

The special Shining
Thicket Rat which
I bought from one
of the children.

Many native mice
are very beautiful,
e.g. the Zebra
Grass Mouse from
savanna grasslands
in northern
Nigeria.

Top: Our adorable Little Mongi, a Cusimanse Mongoose who was always alert, curious and loved digging for insects in our garden.

Bottom: One of our hand-reared Large-spotted Genets sitting on my shoulder.

Moshi who had an amazingly close friendship with an Australian Magpie.

Moshi thought she could do everything my children could do.

Moshi and Cygnet. Moshi expected Cygnet to throw the piece of wood so that she could retrieve it.

Zomba Plateau as seen from the road between Zomba and Liwonde.

Zomba Plateau in the wet season, where David studied the ecology of small mammals in the mosaic of montane grassland and montane forest.

David working in his tent on Zomba Plateau.

David setting
a live-trap in
montane grassland
on Zomba Plateau.

A Nyika African
Climbing Mouse
which lives in
montane grasslands
on some mountains
in Malawi. Its semi-
prehensile tail and
specially adapted
paws and feet
enable it to climb
grasses to obtain
seeds for its food.

Chieffy, the
Gambian Giant
Pouched Rat,
who became very
friendly when I
started treating her
injured foot.

An ephemeral pool in Liwonde National Park, early in the wet season when the trees were bursting into leaf and the grass was beginning to grow again after the hot dry season.

David, with an armed game scout, weighing a mouse in Liwonde National Park in the dry season.

David and a game scout measuring the height and density of grass in Lengwe National Park in the wet season.

Beautiful female Nyala drinking from a waterhole in Lengwe National Park during the dry season.

A Warthog wallowing in glorious mud at the waterhole in Lengwe National Park.

The special noseleaf on the face of a horseshoe bat which acts like a parabolic reflector and focusses echolocation calls into a beam.

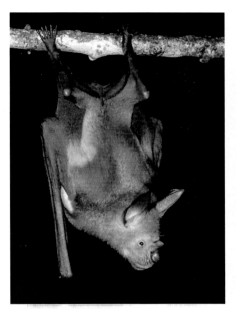

A Striped Leaf-nosed Bat who became tame after being handled gently when she was caught in a mist-net.

One of Africa's most spectacular bats - an African Trident Bat with trident-like projections on its noseleaf and fur that looks like burnished copper and gold.

David photographing Charlie the bull elephant in the game scouts' village at Kasungu National Park.

So confident with a human! Lena feeding banana to a Wahlberg's Epauletted Fruit Bat.

Zebras in the early morning on the Nyika Plateau.

The rare Long-tailed Pouched Rat, *Beamys hindei* – a docile and delightful little fellow.

The sheer ramparts of Mount Mulanje, the highest mountain in Central Africa, as seen from a tea plantation near Likabula.

David with three
of the 17 porters
who carried our
katundu (boxes of
traps and bags of
other equipment)
up and then down
Mount Mulanje.

Ploughed fields
viewed from a
wooded hillside
on Kapalasa Farm.
This farm played
an important role
in the conservation
of wildlife in the
Shire Highlands of
Malawi.

David and I placing
a modified butterfly
net over a furled
banana leaf to
census the Banana
Pipistrelles who
roosted in the leaf.

Just minutes after capture, this Wahlberg's Epauletted Fruit Bat is already trusting me enough to eat a banana while being caressed.

An Angolan Free-tailed Bat snuggling into my hand to find warmth and comfort prior to being released on a cool morning,

Another spectacular bat – the apparently rare *Myotis welwitschii* which has orange and black wings and a black-spotted orange face.

Bilbo, the Silvery Mole-rat, who became very friendly while in captivity and seemed to love being cuddled by my children.

An elephant in the narrow gap between our camp-table and a tree, scratching himself on the tree-trunk. Our tent is on the other side of the tree, behind the trailer.

One of the kanga-babies trying to get into my polishing rag because, to her, it looked like a pouch.

Sallee the kangaroo with Moshi and Nyika waiting for me to return from a visit to the neighbours next door.

Woo, my rescued and hand-raised Crested Pigeon taking his post-prandial nap on my head.

Our much-loved Blossom Possum (R) with one of her daughters having a 'midday feast' in the shed where they slept during the day.

in my hand. By this time, Ruff's excitement was as obvious as a dog's when it knows its dinner is coming! He was bouncing up and down on his forearms, echolocating like mad, and flickering his ears to and fro to pick up the echoes. He snatched a termite and gobbled it down, and then wanted more and more. It was not until his tummy was bulging that he flew out of the door and away and, sadly, I did not see him again. How did he know to go back to the net to attract my attention again? Observations like this suggest that the cognitive abilities of bats have been *greatly* underestimated.

We became very fond of these truly cute little horseshoe bats. Although they have grotesque noseleafs, they have bright little black eyes and large leaf-shaped ears that flicker to and fro as they listen to the echoes of their echolocation calls. They have very round bodies and longish fluffy fur, so they made us think of flying powder-puffs – hence the names Puff and Ruff. On our last day in Zomba, these little bats staged an unforgettable farewell event! I had netted eight of them the previous night and, because we had visitors, I fed them well and settled them in comfortable cloth bags, and I did not examine them until the next morning. By then they were torpid (as they always are during the day in cool weather), but they aroused themselves while being handled and hungrily ate the last of our frozen termites. Then I decided to let them spend the day hanging in one of the very shady cypresses just outside the house – we never discovered where they normally roosted during the day. When I saw them all settled, I packed up all the bat notebooks and equipment and by then I was feeling so tired that I lay down on our bed for a short rest. A few minutes later, one of the bats flew into the room, fluttered over the bed and then flew out of the room. A minute later, two more did the same thing, and then groups of three or more until all eight were flying around the bedroom end of the house. Then they all landed on a lampshade in the hall – the darkest room in the house – and they stayed there all day, looking like a very special decorative fringe around the rim of the lampshade! We were frantically busy that day. We had people coming to read the meters, to collect borrowed goods, and to check the condition of the house and its inventory. We had to finalize the packing up of our personal things, and we were inundated by friends who came to say goodbye. And everyone wondered at the fringe of bats. Then our friends came back with their friends who also wanted to see the fringe of bats! And the bats did not seem to be the least bit concerned.

They stayed wide-awake and flickered their ears at everyone, and everyone was amazed and delighted by them.

Encounters with other amazing bats in Zomba

Those bats flew off at dusk and the next morning we set off for Lilongwe on our way back to Australia. But, long before that, we had many other interesting encounters with bats. My first experience with the largest of the insectivorous bats reinforced my conviction that animals are much easier to handle if their first experience with humans is favourable. The bat was a *Macronycteris vittatus* (Striped Leaf-nosed Bat), a species that has a wingspan of up to sixty-six centimetres, and long, sharp canines easily able to pierce the bodies of very hard-shelled beetles. I saw this bat fly into the bottom of one of my nets, but she was not properly caught and, because it was very important to catch her, I flung myself at her and grabbed her in my thick leather gloves. She put up quite a fight and I was very glad of those gloves, but I eventually had her safely secured in a cloth bag. We needed to keep this bat in captivity for ten days to record how much she ate and how often she drank, so she was released in the spare room – the 'battery' – along with other bats that were being kept for the same reason. But this one never forgot her rough handling and never lost her fear of humans. She was always aggressive even though her only source of food and water was from our hands. We wondered if all bats of this species were naturally very aggressive, but this is not so. Later on, I caught five more of these bats and was able to extricate them gently from the nets and then feed them some termites almost immediately, and all of them became very tame.

Another interesting encounter was with an Egyptian Slit-faced Bat (*Nycteris thebaica*). I was able to extricate this very delicate, long-legged bat from the net without any trouble and I gave it some termites before examining and measuring it, and then I gave it another feed before taking it outside to let it go. It hung from my hand for a few moments and then flew away but, having flown only about twenty metres, it turned around and flew back to me and landed with its tiny toes clinging to my lower lip! It was so light I hardly felt it. Unfortunately, I had no more termites with me, so I just stayed very still and it soon flew away again. If I believed in reincarnation, I would wish to come back as a bat so I could understand what goes on in their amazing minds!

Banana Pipistrelles – the subject of another major project

Our second bat project in Zomba focused on the natural history of *Pipistrellus nanus*, the Banana Pipistrelle (or Banana Bat), so called because these tiny bats roost in the furled new leaves of banana plants and they occur almost everywhere in Africa where bananas grow. Bananas were introduced into Africa about two thousand years ago but, before that, there were other related plants such bird-of-paradise plants (*Strelitzia* spp.) and false bananas (*Ensete* spp.) that also have furled leaves. This means that Banana Pipistrelles have had a very long time to become adapted to using this special roost. They even have suction pads on their wrists to help them climb up slippery leaves! But these bats have a serious problem: banana leaves unfurl after only one to three days, so the occupants have to keep moving from one roost to another – even when they have flightless young that have to be carried to the new roosts. A study in temperate South Africa, on *Pipistrellus nanus* roosting in strelizias, had shown that furled leaves were sometimes occupied by single bats and sometimes by pairs or larger groups. This study also showed that there were equal numbers of male and female adults in the population although the sex ratio at birth was biased in favour of females. The females were monoestrous (one litter/year) and had either one or two babies in November to early December at the beginning of the wet summer season in South Africa. We wanted to find out if the biology of *Pipistrellus nanus* in tropical Malawi was different, and we wanted to find out more about their furled-leaf domiciles.

We decided to visit the nearby village of Malonje, which, like most Malawian villages, had several groves of banana plants nearby. We asked the senior resident for permission to catch bats in the furled leaves, and then enlisted the help of some of the village boys. They were very good at folding the tops of the leaves so they could then catch the bats inside and bring them to me. At first, the boys did this by themselves but, from early December onwards, I went with two of them and we caught the bats in a long net on a long pole that could be placed right over a leaf so that every bat could be captured. In this way, I could determine group-composition in terms of the ages and sexes of the bats in each leaf. I also measured the diameter of the entrances of occupied leaves and discovered that *Pipistrellus nanus* only roosts in leaves with entrance diameters of seven to twenty-four centimetres, and I found out that leaves remain suitable for occupation for only one to three days. I followed the reproduction of these bats and noted that the testes of the males were scrotal from April to October, but they were largest in May. This suggested that mating took place in May, but could this possibly mean that the gestation period of this tiny bat was as long as six months? We were not able to answer this question until we returned to Malawi in 1993-94 for our second trip. The study of the bats from Malonje raised other questions too, and many of these were also answered during our second trip (Chapter 11).

Little Free-tailed Bats – our third project in Zomba

The third project we started in Zomba was on the reproductive biology of the Little Free-tailed Bat (*Tadarida pumila*). Free-tailed bats (family Molossidae) are the mammalian ecological equivalents of swifts: they have very long, narrow wings and they hunt insects that are flying in open spaces high above the ground and the trees. They are not very manoeuvrable, they cannot fly slowly and they cannot take off from the ground. Because of this, they have to roost in places that are high enough to enable them to dive downwards to gain sufficient speed for flight. The Bone family had told us about bats that dived from a crack under the eaves of their lovely old colonial-style house, and they invited us to study them. They knew exactly where the bats dived so, in late October, we set a mist-net across the flight path and caught eighteen of them, most of which were released later the same night. Of these, fourteen were pregnant females. Subsequently, we followed the reproductive biology of the bats at this house at monthly intervals, and we also obtained samples of *Tadarida pumila* from other houses in Zomba. From this study, we found out that these bats had only one baby at a time, that the gestation period was about sixty days, and that all of the females had a litter in the last week of November followed by a post-partum mating that resulted in a second litter being born at the end of January. This birth also was followed by a post-partum mating, and a third litter was born at the end of March. All of these births took place in the wet season when insects were most abundant, but just a few females went on to have a fourth litter at the end of May in the dry season. Subsequently, we compared the reproductive strategies of *Tadarida pumila* in Malawi, Kenya, Uganda and Ghana. The pattern of rainfall in each of these parts of Africa is very different but, in each place, the timing of births was closely associated with peaks in rainfall and the concomitant peaks in the abundance of insects.

For the first four months of this study, the bats in the Bone family's roof always emerged just after dark from the same hole, and always dived almost to ground level before flying away. The bats were always released the same night and, as most were marked, we knew that many of the bats were recaptured several times. However, in February, some of the bats emerged much earlier than usual – before we had set the net – and others made much shallower dives than usual and managed to fly over the net. We had to go back the next night with an extra net that could be set over the top of the usual one. In March, we set the nets extra early, and again set one net above another, and so we obtained a useful sample of the bats, but in April they tricked us again by emerging from a different hole. And in May they emerged from yet another hole! It seemed that those bats that were caught several times were adopting avoidance tactics, and that was not very surprising because they are intelligent animals. But it was not just a few bats that were adopting the avoidance tactics – it was the whole colony – including the young ones as they came along! Who organized them? And how did they tell each other what strategy to adopt?

Bats in the 'battery'

A fourth project that we started in Zomba was extended to cover all the species that we caught at all localities, and it was motivated in part by our desire to make the best possible use of the specimens that we had to keep as vouchers or donate to the collections in Malawi. This project was to study the structure of the kidneys (renal form) and their function. In bats, the extent to which the kidney can produce concentrated urine (as a means of conserving water) can be accurately predicted from the form of the kidney. We already knew that in rodents, species that live in desert and semi-desert habitats where water is scarce, have more efficient kidneys than species from wetter habitats. Did this apply to bats too? Luck favoured us with this project. We were able to feed the captive bats on the winged termites we had caught in Lengwe National Park and bats that we captured ate them with great enthusiasm because they are rich in fat and very nutritious. For the renal form and function study, we kept twenty-four species in captivity in the spare room of our house – the 'battery'– where they could fly at night and roost in a variety of places such as under sheets of bark on the bed, in folds of the curtains and in the foliage of small branches of trees. The bats were hand-fed at dawn and dusk and were offered water as drops suspended from our fingers or in teaspoons. We – and this included the Haplets – recorded how many termites were eaten by each individual and whether or not it drank water when it was offered. Twelve species drank every time water was offered, four species were irregular drinkers that sometimes went three or four days without drinking, and six other species did not drink at all! The non-drinkers included female *Tadarida condylura* and *Tadarida pumila* who were pregnant, and one *Tadarida pumila* who was lactating while raising her baby in captivity. Eventually, we were able to compare the renal form of each species with its drinking behaviour in captivity, its diet, its foraging behaviour, flight characteristics, wing-shape and where it roosted during the daytime. We hypothesized that the ability of a bat to conserve water by producing concentrated urine would be advantageous if it lived in an arid environment, if it foraged by flying long distances at high speed (when evaporative water loss from the wings would be very high), if it roosted in very hot and dry domiciles (such as under corrugated-iron roofs), if it roosted in solitude as opposed to huddling together in tight clusters, and if it lived at high altitudes where low temperatures and shortages of insects meant that it had go into torpor for several days at a time, or go into hibernation. These hypotheses were tested and we concluded that, while the ability to conserve urinary water may have been an advantage in most of these situations, it was not essential except perhaps in species that habitually went into longish periods of torpor or hibernation.

The Haplets did much of the feeding and they kept very careful records. Most of the bats became very tame and it was easy to approach them to feed them. Two species, *Taphozous mauritianus* and *Scotophilus dinganii*, even learned to leave their roosts and come when they heard us clicking our fingers or rubbing termites in our hands to warm them. The *Taphozous* greatly amused us by the way they scuttled sideways over the bed by bouncing on their feet and wrists. They looked just like mechanical toys and Jonathan was sure that they should have had keys sticking out of their backs!

Keeping all these bats in captivity enabled us to observe their grooming behaviour, some of their social behaviour, and how well they could fly in the 3.5 × 3.5 × 2.5 metre 'battery' or in a corridor that was only one metre wide. But the most important advantage of having the captive bats was that the Haplets and all their friends learned to love them and realize that bats are not the fearful creatures of ghost stories and darkness, but warm, cuddly, friendly, intelligent and thoroughly delightful little animals.

Working on bats at Liwonde National Park – it was fraught with adventures!

We started setting mist-nets in the vicinity of the game scouts' houses near the Administration Camp. Everyone agreed that we were not likely to be bothered by lions and elephants so close to where people lived, but there was one night in November when lions in the vicinity did prevent us opening the nets we had already hung up and, on another night, a group of Bushpigs came into the village and had me running for the shelter of our house! We always wore powerful head-torches on our heads and always looked around for the shining eyes of any large mammals. We sometimes spotted eyes shining in the vegetation at the edge of the village clearing, but they were usually those of African Civets, which were no threat to us. And eyes that were like bright red lights swinging around in the trees were those of bushbabies, in this case Southern Lesser Galagos.

But I had an adventure one night! I spotted a pair of eyes watching me from the edge of the clearing. I had set a mist-net nearby and I had hung some cloth bags containing newly captured bats on the poles supporting the net. The idea was that the bats in the bags might call and attract other bats into the net but, instead, the calls appeared to be attracting a small carnivore. I wanted to see what it was, so I started making a squeaking noise between my lips and a finger – as birdwatchers do in Australia to attract birds. The carnivore was very interested and it emerged into the clearing. It was a Large-spotted Genet, now considered to be the same species as those we had hand-reared in Nigeria. I continued squeaking and the genet crept slowly closer and closer until it reached the base of a sapling where I was standing. Then it climbed the sapling until it was level with my head but, when it crouched ready to jump onto my shoulder, I decided discretion was called for and I moved out of reach! The poor genet promptly panicked and raced for the bush.

During our nine visits to Liwonde National Park, we caught nineteen species of bats, including six that we never caught in Zomba, and most were new records for the park. Three of these were microbats that we began to think of as 'Liwonde specials':

Myotis welwitschii, a 'rarely-recorded' species that has orange dorsal pelage, cream ventral pelage, orange ears with black spots, and spectacular wings with some areas black and other areas orange with black spots.

Pipistrellus stanleyi, a new record for Malawi.

Pipistrellus zuluensis, in Malawi known only from Liwonde National Park.

Prior to our trip to Malawi, David Harrison, a friend of ours, had set mist-nets over the Likwenu River that we crossed on our way from the entrance of the park to the Administration Camp, and he obtained the first record of *Pipistrellus zuluensis* in Malawi from that locality. We were keen to set nets across the river too. In the wet season, the Likwenu is a raging torrent but, by the end of the dry season in October, it is reduced to a series of pools and there was a promising one on the west side of the bridge. We knew we would need protection if we went to the river at night, so we took game scout Joseph with us, and his .303 rifle gave us a comforting sense of security. We set two nets close to the edge of the pool and waited. And then darkness fell and the atmosphere changed. We caught two bats and saw many others managing to avoid the nets as they flew over the pool to drink. The bats held our attention for a while, but then we became aware that Joseph was far from happy. He kept looking over his shoulder, and he jumped at every sound. I said, "Joseph, I think that you are not happy about us staying here by the river. Is something worrying you?" He looked somewhat embarrassed but then confessed, "Well, Madam, there is a big pool under the trees on the other side of the bridge and there used to be a very big crocodile living in it." We packed up and went home even though it was only half past seven!

In November, we drove along the track that circumnavigates 'Happold's Hill' and noticed an enormous baobab tree with a large slit in its trunk. The largest of the African slit-faced bats, *Nycteris grandis*, had been found in hollow baobabs near Liwonde township, so we wanted to check our tree to see if there were any of these bats roosting inside it. We had an armed scout with us and he warned us to approach very carefully in case the hollow trunk was occupied by a leopard! We did indeed approach carefully, but no leopard jumped out, so I climbed up and looked inside. There were no bats in sight but, lying on the ground at the bottom of the hollow, there were three human skulls with their skeletons buried in debris! My first thought was that the hollow was the den of a man-eating leopard, but the scout assured us that this was not likely. He did not offer any explanation for the skulls so we reported the find to the chief game warden and asked him if we should inform the police, but he said there was no need. The track to that baobab became impassable in the wet season, but we went back to it in June, on our last trip to Liwonde. This time there was a colony of seven *Nycteris grandis* in the hollow trunk and we caught three that we brought back to Zomba with us for observation. This bat is the only species in Africa that eats small vertebrates, including frogs, fish, birds and small bats, as well as insects. Back in Zomba, David had to save the skins and skulls of some mice for museum collections, and I offered the skinned bodies to the bats. They wolfed them down, bones and all! And they could easily fly around the room with a mouse in their jaws.

There was another species of slit-faced bat in Liwonde that was also of great interest. This was *Nycteris macrotis*, a smaller species that roosted in the culverts under the main road through the park. We made our first discovery of these bats in a culvert not far from the Administration Camp in early February. We were very keen to catch the single occupant of this culvert, so we put a net across one end and then "darling husband crawled through the culvert, which was full of sand and water." That is a quote from David's record of the event in our bat field notebook! And we did catch the bat! Throughout all this, our game scout Joseph looked on and he was speechless when David crawled out, covered with sand and mud from head to foot. We were not surprised when he suggested that David should walk back to the game scouts' village instead of going in the Kombi, and he assured David that he could safely walk back along a path that several people had already used that morning without seeing any lions or elephants. The odd thing was that neither of us stopped to wonder why Joseph did not walk back with David. Instead, he hopped into the Kombi with me and we drove to the village. Then Joseph jumped out and ran from house to house until the entire community was out waiting to see the unbelievable spectacle of a mad Englishman emerging from the bush, covered in mud from head to foot!

After that, the scouts suggested that we should employ some of the young boys to crawl through the culverts, and we were very happy to oblige because the families needed any extra money they could get. However, this put our Jonathan's nose out-of-joint! Eventually, in June when all the culverts were dry, we said he could have his turn. We put a net across one end of a culvert that had quite a lot of bats hanging from its roof, and Jonathan crawled in at the other end. He flushed several bats out and into the net, but others flew back over his shoulders and settled behind him, so he had to go through again and then a third time. Each time he emerged, he had to brush past a clump of grasses and it looked as though he were collecting grass seeds as he did so. However, as soon as we started to brush the seeds off, we realized that they could move! They were ticks! Poor Jonathan! He was stripped and made to stand on the bare road while we removed every tick. Fortunately, there was not enough time for them to have become embedded and that was a great relief because African ticks can carry nasty diseases. The game scouts told us later that Jonathan's ticks came off warthogs that often run into culverts to escape from lions. This adventure had an extra happy ending. On our last night in Malawi, David gave a talk about our work to the Lilongwe branch of the Wildlife Society of Malawi and, for a bit of light relief, he told the story of Jonathan and the warthog ticks. At the end of the talk, an English girl who had just arrived in Malawi told us that her mission was to collect warthog ticks and take them back to England to find out if warthogs were a reservoir for swine fever, a disease that could be transmitted to domestic pigs. She thought this would be a difficult mission, but we were able to tell her exactly where to go to obtain the ticks. She did, and subsequently found out that they did not carry swine fever and that pleased everyone who wanted warthogs to be protected without controversy.

Our sad last day in Liwonde National Park

After living in the game scouts' village for a week every month for almost a year, we felt very sad to be leaving this friendly and generous community. They had welcomed us so warmly and made us feel so much at home with them. The Haplets were as sad as we were and they brought gifts to give to their playmates during our last visit. Lena had spent weeks sewing thirty little dolls for the girls, and Jonathan gave the boys a real leather football and pump. In return, and especially to thank Jonathan for playing his recorder to the children, the village community planned to put on a concert for us on our last evening.

We had been out in the bush, doing some last-minute checks, and we walked back to the village feeling very excited about the concert. However, as we came closer, we heard all the ladies keening and we realized that someone had died. It was the wife of one of the game scouts, who had been in hospital in Zomba and she had died there. It was about four o'clock when the keening started, and the ladies kept it up until dusk. It was heart-wrenching! We kept out of the way, but we were aware that the men were collecting large bundles of firewood from the bush and getting it ready for a large fire in the clearing. Obviously, there would be no concert, but what actually happened that night was unforgettable. As soon as it got dark, the fire was lit, the villagers gathered around it, and then they began to sing! We had already heard groups of Malawians singing together in harmony and with counterpoint – the blending together of different tunes. But this surpassed anything we had ever heard in Malawi, and it was as exquisite as any singing we had ever heard anywhere! We think there were about eight main singers and they included Joseph and one other scout who had deep basso profundo voices and sang as wonderfully as Paul Robeson. Sometimes, these men just hummed an accompaniment to the songs, and then it was like an organ playing. None of the singers had had professional training but they had the sweetest voices and singing in harmony and counterpoint is something that seems to come almost naturally to Malawians. Some of the songs were hymns. Others were madrigal-like songs that were probably purely Malawian in origin. We assumed that the singing was to speed the soul of their departed friend on its journey. It was kept up all night long and it was so beautiful that David and I could not bear to go to sleep. It was something we shall never forget.

Working on bats at Lengwe National Park – not quite so adventurous!

At Lengwe National Park, we always set a net across the flightpath of the free-tailed bats as they dived from holes under the corrugated-iron roof of the toilet-block near the visitors' chalets. These were mostly Angolan Free-tailed Bats (*Tadarida condylura*) the largest of the species in this family that we caught in Malawi. They commonly roost under the roofs of houses and are considered pests because they become very noisy in the late afternoon. Furthermore, they produce copious amounts of guano and their roosts are always very smelly. The intense, all-pervading musky smell of the bats in the toilet-block is one of our lasting memories of Lengwe, but these bats must do an enormous lot of good by feeding on the moths of the cotton bollworm, which is a greater pest than the bats are. A lot of cotton is grown in the Lower Shire Valley.

The first night that I set a net near the toilet-block was almost a disaster! I suddenly found myself having to deal with a net about to be filled with potentially hundreds of furiously biting and struggling bats that are particularly good at tying mist-nets into knots and then biting them into countless holes! David, at the time, was cooking supper, but I shouted for help and he and the Haplets helped me to move the net out of the flight path. Some of the bats managed to free themselves and fall to the ground, but they could not take off from the ground and we managed to catch a lot of them before they could scuttle away. Then we disentangled the others. Altogether, we caught twenty-one bats in just a few minutes.

It seemed well worthwhile to catch these bats every time we visited Lengwe National Park so that we could follow their reproductive biology and compare it with that of the *Tadarida pumila* that we were studying in Zomba. We found that there were some months when the catch from the toilet-block was too small to be useful, so we started to set nets to catch the same species as they emerged from roosts under the houses of the game scouts in a village a few kilometres away and, in this way, we obtained useful samples every month. These bats were not easy to handle in the nets and we had to wear heavy leather gloves to prevent them biting us. However, they quickly settled down in their cloth bags and, very often, I did not check them until the following morning by which time they were usually torpid. I could handle them then without gloves and they seemed to love the warmth of my fingers. If I half-closed my hands, they would push themselves backwards into the 'tunnel' and they no longer attempted to bite. It was not easy letting these bats go. They had to be warmed up and then thrown high into the air so they could dive to gain sufficient speed for flight. Alternatively, we put them onto the trunk of a tree so they could climb high before diving off and flying back to their roost. Free-tailed bats are very well adapted for scuttling and climbing.

Tadarida condylura at Lengwe had only one young per litter and each female had two litters per year, as in all other parts of Africa. In Kenya, each of the two litters is born in one of the two wet seasons, but the wet seasons are in November-December and in March-April, so there is a gap of about three months between the consecutive litters. In contrast, we observed that the bats in Malawi had to squeeze the two litters into the one wet season between November and March, and therefore there had to be a post-partum oestrus and mating. Ours was the first record of post-partum oestrus in this species. In Uganda, where there is rainfall in all months but with peaks of heavy rain in March-May and in August-November, the births occurred in February-March and in July-August, just before each peak in rainfall. This meant that the gap between litters was about four months. It is amazing how the reproductive chronology of a single species can be adapted to suit the prevailing climatic regimes so that the females are always lactating during periods of peak rainfall when insects are most abundant.

As well as regularly setting a net by the toilet-block, we set other nets in the clearings around the chalets and, once only, across the road that ran through riverine forest on the way to the entrance of the park. On that occasion, I ventured off the road in search of suitable clearings or flyways where we could set nets in the forest. And there I came across the most enormous snake I have ever seen! It was greyish in colour and about three metres long. I didn't stop to pass the time of day with this fellow, and David and I hurriedly moved the nets further down the road. I will always wonder if that snake were a deadly Black Mamba!

In total, we caught only ten species of bats at Lengwe National Park, and all were new records for the park, but not for the Lower Shire Valley. Undoubtedly many more species inhabit Lengwe National Park.

Chieffy, the Gambian Giant Pouched Rat

During the weeks that we spent in Zomba, we had time to enjoy some very unusual pets. The word soon got around Zomba and the neighbouring villages that zoologists had come to town, and some of the local young men hoped to cash in on this by bringing us animals and birds that we might buy from them. We strongly discouraged this and always confiscated the animals – but animals came just the same. One animal that was brought to us was a Gambian Giant Pouched Rat or Chief Rat (as many people call them), one of the largest of the African rats. The one that came to us was an adult female. She had a head-body length of about thirty centimetres, a white-tipped tail that was a little longer, and she weighed about 800 grams. Chief

Rats, if handled early enough, make very nice pets but wild-caught adults have a fearsome reputation for aggressive biting! Therefore, the young men had tied her wrists and her ankles together with coarse twine, and they carried her, dangling belly-up, underneath a stick. I was furious because the twine had been tied so tightly that circulation to the paws was jeopardized, but consideration for animals is not something most Africans have in their culture. It was impossible to let the Chief Rat go until the damage to her paws had been assessed but our garage had a concrete-floored annex that made an ideal temporary home for her. We quickly covered the floor with soil, logs and vegetation and then cut the bonds and turned the Chief Rat loose. At first, everything looked promising but, after a day or two, one back foot became infected and very swollen, and the poor animal became very lame and conspicuously unwell. We had some antibiotic cream with us, so I cornered her and managed to get some of the cream onto the infected foot: she was, I suspect, too sick to put up any resistance. However, within a day or two, she allowed me to approach her and stroke her head and back, and she would even poke out the sore foot so I could easily put more cream on to it. By this time, she was obviously no longer sick, the foot was healing well and she was no longer limping. Not long after that, she would climb onto my lap and sit there while I stroked her or fed her titbits of peanut butter, honey and maize-meal. She was a lovely animal and I was sad when the time came to let her go – but she deserved her freedom.

Bilbo, the Silvery Mole-rat

Another animal that was brought to us was a Silvery Mole-rat that had been found when fields were being dug up in preparation for growing crops. These very odd-looking rodents live underground like moles, and they dig long tunnels in their search for the roots, bulbs and tubers that they eat. They are rather like sixteen-centimetre-long cylinders with rounded ends, covered by long, silky, silvery-grey fur, but the oddest thing about them is that their long, slightly curved, incisor teeth – two uppers and two lowers – protrude through the lips and are therefore visible at all times! Mole-rats use their incisors like chisels to scrape through hard soil and roots and, by keeping their lips shut, they can do this without getting their mouths full of soil. Mole-rats have short legs, very short tails and minute eyes and they lack external ear pinnae. Our little fellow, who we called Bilbo, was very put out by being dug out of his burrow and dropped into a bucket, and he jerked around making very fierce, scratchy grunts. He clicked those wicked-looking teeth together in a way that left little doubt in our minds that he would use them on us if given half a chance! And we imagined that teeth designed for chomping through hard soil and roots would quite easily chomp all the way through a finger! We bought a large metal bucket in Zomba market and filled it three-quarters full of dampened soil for Bilbo to burrow into, and he stayed 'underground' for most of the time except after dark when he surfaced to search for the Irish potatoes, sweet potatoes and carrots that we put on the surface for him to eat. At first, he always dragged his food into his burrow but, after a week or two, he got used to us watching him and often stayed above ground while feeding. I started getting him used to my hand in the bucket, and then he took food from my fingers and let me stroke the fur on his rump. By that time, he had stopped attempting to bite and had stopped snorting and hissing, so I decided to see what would happen if I put my left little finger in front of him. I thought I could survive without this finger if I had to, but I need not have worried. Ever so gently, Bilbo took my finger between his incisors and tugged my hand as far down his burrow as it would go! It was amazing! After that, we realized that he had become delightfully friendly and he actually seemed to love being picked up and cuddled. The Haplets adored him and played with him for hours. We let Bilbo go in suitable habitat but another one was brought to us some time later, and he too became just as friendly. There is a sequel to this story, but that did not happen until we returned to Malawi in 1993 and visited Ntchisi Mountain (Chapter 11).

Journey to northern Malawi

A particularly useful result of David's study of the rodents on the top of Zomba Plateau was the discovery that the grassland species never went into the patches of montane forest. We knew that there were montane grasslands on the tops of all the other isolated mountains and plateaux in Malawi, and also on the tops of the isolated mountains that form a chain down eastern Africa from Ethiopia to the Drakensburg in South Africa. On all mountains, changes in altitude are associated with changes in climate that result in zonation of the vegetation. Latitude is involved as well – the closer the mountain is to the Equator, the higher one has to go to find particular zones of vegetation. We wondered if all the mountains in Malawi had different species of grassland rodents. If the grasslands had always been isolated, one would expect different species on each mountain and, conversely, if the mountains had the same species, the grasslands must have been linked at some time or times in the past. In eastern Africa, there have been alternating warm-wet (pluvial) periods and cool-dry (inter-pluvial) periods associated with the various ice ages in the northern hemisphere that occurred during the Pleistocene and probably as far back as the Miocene (five to twenty-five million years ago). During the warm-wet periods, forests spread more widely over the lowlands and also spread up to higher altitudes on the mountains – perhaps even to the tops of the mountains? During the cool-dry periods, the forests contracted and perhaps were unable to grow at higher altitudes where, in contrast, montane grasses *could* grow. Perhaps the forests contracted so far off the mountains that montane grasslands were able to link the tops of adjacent mountains? Alternatively, perhaps all the montane grasslands were created entirely by humans burning or cutting down the forests? Professor Olov Hedburg (a famous Swedish botanist) maintained that eighty per cent of the vascular plants in different African alpine habitats are endemic to their particular mountains, suggesting that the alpine habitats on separate mountains have been isolated for long periods of time, and he suggested that colonisations of mountains, and dispersal between mountains, might only have been possible during cool-dry periods. In contrast, another botanist (Professor Michael Meadows from South Africa) thought there was no evidence that the alpine vegetation zones ever descended low enough to allow alpine plants to spread from one mountain to another.

Over the years, the attention of zoologists has focused on changes in the distributions of forest mammals in response to climatic fluctuations, but very little work had been done on grassland mammals – especially those of montane grasslands. For this reason, we wanted to study the distributions and habitat preferences of the small mammals in other montane habitats in Malawi and we hoped this would provide more information about past climates and the distributions of small mammals in the past. We also hoped that this might throw light on the controversies raised by the recent botanical studies mentioned above. Were the mountains ever completely covered by forest? Were the mountains ever completely covered by montane grasslands that were extensive enough to form corridors between adjacent mountains? Or do montane grasslands only exist because of human activities?

These questions intrigued us and we thought we could throw some light on them by comparing the small mammal faunas on Malawi's highest mountains. So, we all set off in mid-April to visit the Nyika and Viphya Plateaux in the north and, en route, we surveyed small mammals and bats in Kasungu National Park on the Central African Plateau. Then, in May, we went to Mount Mulanje in the south. These trips meant that we missed one of our regular visits to Liwonde and Lengwe, but we felt this was justified.

Kasungu National Park – stories to dine out on!

We set off from Zomba at seven o'clock in the morning of Lena's eleventh birthday – 17 April – and drove the familiar route to Liwonde National Park but, instead of turning right at Liwonde township, we crossed the Shire River and continued across the Upper Shire Valley towards the escarpment. The previous time we had crossed this area – on our way from Lilongwe to Zomba when we first arrived – it had been dark. This time we realized that this area (which is the Liwonde Flood Plain) is heavily cultivated, with cotton and maize being the main crops. It is also comparatively densely populated, and we passed many small villages of mud-brick and thatch houses and saw many people walking or cycling along the road. We drove steadily towards the blue escarpment, reached it and climbed up through Balaka and Ntcheu until we reached the flattish Central African Plateau. There was much less cultivation on the plateau – especially on the Mozambique side of the road – and fewer villages, but we saw more cattle and goats. We passed several young lads who were hoping to sell us roasted mice threaded onto sticks. The mice were *Mastomys natalensis*, which live in houses as well as in the wild, and they breed so rapidly that they form plagues in the wet season. They can be serious pests of crops and they carry some very nasty diseases, so we did not disapprove of the mouse-kebabs – but we had no desire to try them!

Towards the east, the panorama is magnificent. Rolling hills extend, ridge after ridge, to the edge of the escarpment and further away we could just see Lake Malawi through the haze. We came to Dedza, a small town at the foot of Dedza Mountain, and stopped there to buy petrol and to have coffee and a special jam pie to celebrate Lena's birthday. Then we drove on to Lilongwe to have *chambo* and chips at the hotel. We also spent a short time looking at curios for sale in a market outside one of the main supermarkets. There were many traders there, each with his own stall. We were particularly interested in building up a collection of carved wooden fish because the fish in Lake Malawi are so extremely diverse in shape and ecology.

After Lilongwe, we drove on over a good tarmac road that crossed rolling downs with valleys and streams and rivers, and we also went through some extensive areas of woodland savanna. There seemed to be very little farming in this area. Eventually, we reached the town of Kasungu and then we had forty-five kilometres of bumpy, sandy track in front of us before we could reach our destination – the Administration Centre of Kasungu National Park. It was getting late and we had to pass through the main gate into the park before dark, but the setting sun backlit the seed-heads of a lovely grass called *Themeda* and made them glow red and pink. It was very beautiful and, of course, we *had* to stop to take some photos!

We only just reached the park before the gate was closed for the night, but then we had to drive on in the dusk and we were worried that we would not be able to find our way to the Administration Centre. We also had to keep looking out for animals because they tend to wander carelessly over the roads at night. At one place, we inadvertently drove through a herd of elephants and I glimpsed one, in silhouette, waving its trunk and flapping its ears very angrily. The angry trumpeting of others was also far too close for comfort and we drove away as fast as we dared!

At last, we reached the Administration Centre and located the house of the resident biologist, Richard Bell, who had invited us to spend our first night in his house. We had supper and were glad to bed down for the night.

The next morning, Richard showed us his Wildlife Research Unit and then took us to see a display at Lifupa Lodge where tourists can stay. Kasungu National Park conserves habitats typical of the Central African Plateau. The terrain is undulating with dambos (seasonally marshy grasslands) along the drainage systems, and isolated rocky inselbergs. Most of the park is covered by miombo woodland dominated by *Brachystegia* and *Julbernardia* trees that lose their leaves in the dry season. Over cool drinks at Lifupa Lodge, Richard explained the research he was doing on the feeding habits of the large herbivores. He had noted that areas that have different fertility supported different species of grasses and were grazed by different species of large herbivores. He wanted us to set traps to find out if the species composition and densities of the small mammals were also influenced by the fertility of the soils. Therefore, over the next five days, we set live-traps in eight different habitats.

At all times, we were accompanied by Mr Moyo, a very knowledgeable Malawian game scout who identified all the footprints we saw on the sandy tracks we walked along. Lion and leopard footprints were common and, needless to say, Mr Moyo was armed with a heavy-gauge rifle and he often brought another armed game scout with him.

We had an amusing encounter very early one morning just as all of us had all got out of our Kombi prior to walking into the bush to check traps. A car pulled up behind the Kombi and an American lady got out. She was looking rather surprised and her first words, spoken with a very pronounced twang, were "My oh my! Can you tell me why you guys are carrying GUNS? GUNS in a NATIONAL PARK?" Mr Moyo gently explained that David and I were about to walk some distance through the bush and he and his fellow game scout were coming to protect us. The American lady's jaw dropped and her eyes came out on stalks. "Oh ma Gawd!" she exclaimed. "D'ya mean to say there are wild animals? In THERE?"

We were lucky. We did not run into any dangerous animals in Kasungu National Park although a herd of several hundred buffaloes trampled over one of our trap-lines one night. They smashed all the vegetation but, miraculously, very few traps were irreparably damaged. The smell everywhere was amazing – rather like a cow-yard – and there was dung everywhere.

Rudi Schenkel and the Black Rhino

Richard Bell had a fund of stories to tell. One was about Rudi Schenkel, an authority on rhinos. According to Rudi, White Rhinos are notoriously unpredictable – sometimes they charge people but sometimes they don't. In contrast, Black Rhinos are *very* predictable – they *always* charge! Rudi was working in a habitat of Black Rhinos and he had an area fenced off so that rhinos and other very large herbivores were excluded. He wanted to measure the effects of this exclusion on the vegetation. One day, he was walking towards the exclosure, in bare feet, and he was charged by a Black Rhino. Nearby, there was a tree with a large branch leaning down and touching the ground on one side. Rudi sprinted to this tree and began to circle round the trunk. *He* could duck under the fallen branch but the rhino had to go around it. For a while, Rudi managed to keep ahead of the rhino but he realized he could not keep going for as long as the angry rhino could, so he made a leap for another horizontal branch, hoping he could tuck his legs up and out of reach. Unfortunately, he couldn't raise his legs high enough, fast enough, and the rhino charged into them and knocked Rudi off. He fell onto the back of the rhino and found himself astride the then stampeding animal. He knew he could not stay astride, so he tumbled off over the rhino's tail and landed on his back in the grass. And of course, the very predictable Black Rhino turned and charged again! Rudi put his feet up in a last-ditch attempt to ward the rhino off, but then the very *unpredictable* happened. The rhino cannoned into his feet, stopped, and then calmly walked away! According to Richard Bell, the thing Rudi remembered most clearly was how velvety soft the rhino's lips felt on the soles of his bare feet!

Charlie, the elephant at Kasungu National Park

Our visit to Kasungu National Park seemed a golden opportunity to conduct a short survey of the park's bats, but Richard was very negative about this. He said it was far too dangerous to work in the park after dark. I tried hard to persuade him to let me set nets in the garden around his house, bearing in mind that the house would provide shelter if needed. But Richard said that elephants often wandered through the garden, and there had been a leopard close to the house "only last Sunday". I told him that we were used to working in Liwonde National Park where there were lions and very tetchy elephants, and in Lengwe National Park where there were buffaloes, and I said that we always wore powerful head-torches that would pick up the eyeshine of any animals. And, eventually, he was persuaded to let us set nets close to his house. Of course, I always checked for eyeshine before leaving the house to check nets, and Richard usually came with me, but I didn't really think that anything would venture into the garden while people were moving around with bright lights – and nothing did on the first night. However, just as we had unfurled the nets at dusk on the second day, an enormous bull elephant strolled into the garden. He walked past the Kombi and towered over it! Then he went across to the Bells' chicken yard, dropped his trunk over the high fence, and sucked up all the chickens' food. And then he started walking straight towards the one and only mist-net in pristine condition that we had kept especially for Kasungu National Park and the Nyika Plateau! All our other nets had holes in them. I told Richard that the loss of this net would be a disaster and asked him if there was *anything* we could do to make the elephant go somewhere else. Richard was marvellous. He said, "I know this elephant. If we stand quietly in front of the net, he probably won't walk through it." So we did.

It was an unforgettable encounter! Somehow I knew that this elephant meant us no harm. He approached until he was eight metres away and then stopped and quietly watched us. His trunk waved towards us to check our scent and his ears fanned gently to and fro. It was magical! We watched each other for several minutes and then Richard said, "If we move forward slowly, I think we'll be able to chivvy him around the side of the net." So we did, saying to him, "*Choka! Choka!*" It seemed perfectly natural to talk to him in Chichewa, and indeed he *did* walk quietly away around the side of the precious net.

This elephant – known as Charlie – was well known to the game scouts. He was a solitary bachelor male who was often seen in the Administration Area. We had two further encounters with him. The first was one night when we had gone to the game scouts' village to set a mist-net to catch "very small bats" that were regularly seen leaving their roost under the roof of one of the houses. Mist-netting was something the villagers had never experienced, and all the village children were allowed to stay up and watch. The bats turned out to be *Nycticeinops schlieffeni* and we netted the first group that emerged. Then, suddenly, all the children ran for cover shouting, "*Njovu! Njovu!*" ("Elephant! Elephant!"). Two of the older boys were very quick off the mark – they pulled the poles supporting the net out of the ground and, carefully keeping the net taut, they ran with it into a narrow gap between two houses. We went with them and watched as Charlie strode through the village on his way to the rubbish dump where he hoped to find something nice to eat. As soon as he had passed, the boys took the net back to its original position, and we caught some more bats. Then the circus was repeated when Charlie returned from the dump.

That same night, we set traps all around the village and in some of the houses because we wanted to know what species of rodents were living in close proximity to the human dwellings. We went out at dawn the next day to check these traps. It was an amazing morning. It was slightly misty, the sky was golden and the first rays of the sun gilded the tall, dew-bespangled grasses. We took some photos and I said to David, "Wouldn't it be just perfect if that elephant were standing just there!" And a moment later he was! We photographed him as a dark silhouette against the golden background.

Soon after we finished checking the traps, Charlie wandered into the village again but, this time, he was obviously in a bad mood. We noticed that he had a festering gash on one cheek that was undoubtedly painful. Everyone treated him with *great* respect that morning – especially after he had made some mock charges. However, by keeping close to the houses, we obtained some super photos. He was so close that we had to use a wide-angle setting to get all of him into the frame.

We recorded eight species of bats in Kasungu National Park and all were new records for this park. One was *Pipistrellus stanleyi*, a species that we had recorded at Liwonde National Park as a new record for Malawi. We also identified a specimen of *Myotis welwitschii* – the spectacular black and orange vesper bat – in a small collection of specimens at the Wildlife Research Unit. Our catch undoubtedly represented only a small fraction of the bats likely to occur in this national park – we knew, for example, that two Dutch bat biologists had recorded four species of fruit bats when trees in the park were bearing fruit.

The wild and remote Nyika Plateau

On 27 April, we set off for Nyika National Park, which conserves the highest and most extensive plateau in central Africa and much of its surrounding escarpment. Most of the plateau lies between 1800 and 2400 metres above sea level, but its highest peak rises to 2607 metres. Almost all of the plateau is undulating montane grassland, but there are isolated patches of montane evergreen forest in the folds of the hills, and there are montane swamps along some of the drainage systems. There are also some areas of forest dominated by the native conifer *Juniperus procera*, and around the forestry administration centre and visitors' chalets there are mature plantations of exotic pine trees and eucalypts. The plantations were laid out by members of the Commonwealth Development Corporation who envisaged the Nyika Plateau as a vast source of timber and also a vast cattle ranch, but no one had stopped to consider the insurmountable problem of transporting the timber and beef away from that *very* inaccessible high plateau! The climate of the Nyika Plateau is dramatically different from that of the surrounding country. Although the rainfall is not unusually high, the plateau is often cloudy and cool, and even frosts occur occasionally in the coolest months.

Our journey from Kasungu as far as Rumphi was easy although we did have a puncture and dozens of children materialized, like a conjuring trick, from nowhere and gathered around to enjoy the spectacle of a tyre being changed by two whites and their children. But, after Rumphi, it was a nightmare! We had to turn off the main road and soon discovered that a bridge had been washed away in the wet season, which forced us to make a long detour along a track that was almost impassably rutted and made worse by very recent rain. To this day I don't know how we made it back to the road we should have been on. This road up the escarpment, thank goodness, was either gravel or sandy and it was reasonably well drained.

We came to the entrance of Nyika National Park while we were still in the belt of miombo woodland that surrounds the plateau. We introduced ourselves to some of the staff, and then went on. The road climbed steadily and we drove through a zone of proteas and grasses until we reached an altitude of about two thousand metres when the true montane plateau landscape unfolded. David was ecstatically happy. He had known about the Nyika Plateau for a long time and, like almost everyone who comes to Malawi, he was longing to get there. However, it is a very remote place and access is only possible in the dry season – and, even then, it is not easy. We thought we were very lucky and privileged to get there.

The plateau landscape reminded us of Dartmoor in England except that there are no tors or rocky outcrops on the Nyika Plateau. The grasses were short and brown with dark shadows under the wonderful, sometimes stormy-grey clouds that drifted overhead. The grasslands spread over rolling hills as far as the eye could see, but there were patches of bracken here and there, and dark green forests in the folds. It was a landscape that did not seem to be African until we saw herds of Crawshay's Zebra, Southern Reedbuck and Common Eland, and also Roan Antelope whose black-and-white patterned faces and large, beige ears with floppy tips made them look just like court jesters. All of these animals were amazingly tame and often let us walk as close to them as ten metres.

Our first destination on the Nyika Plateau was the forestry settlement at Chelinda where there are cabins for tourists, houses for some of the park officials, and the home of the resident biologist who happened to be an old friend from our days in Nigeria. We had about sixty kilometres of soggy, muddy road to follow and it was not easy to negotiate because graders had been scraping the road into ridges prior to smoothing away the ravages of the wet season. We were very tired and very cold by the time we arrived at Chelinda just as it was becoming dark. However, our cabin was equipped with a Rhodesian boiler – a forty-four-gallon drum of water mounted above a fire – and this enabled us to have hot baths before we walked through the pine forest to the home of our friends, Mike Dyer and Sigrid Johnson. Sigi had prepared a delicious meal – mainly concocted from tinned corned beef since supplies to the Nyika Plateau were very limited – and we ate it in front of a log fire. Then, after much talking and reminiscing, we drank mugs of hot milk spiced with nutmeg, cinnamon and rum, and then walked back through the trees to our cabin. We had most of our evening meals with Mike and Sigi, and sometimes our walks back to our cabin were made very spooky by the weird rising 'whoops' of Spotted Hyaenas who were just a trifle too close for comfort. We hoped they were not feeling hungry!

Our first priority on the Nyika Plateau was to set lines of twenty-five live-traps in montane grassland and montane forest habitats for three consecutive nights. Then, based on the results, we set up two grids of fifty traps just as we had done on Zomba Plateau.

It was wonderful to be able to work freely on the Nyika Plateau without having to worry about lions, elephants and buffaloes. There were leopards up there, but the locals did not consider them dangerous unless they were cornered, so we didn't either. We sometimes heard them calling – it sounds a little like someone sawing wood – and once we spotted one watching us while we were checking traps, but it was too far away to see very clearly. As we walked over the hills to our grids, we often saw herds of zebra, reedbuck and Roan Antelope, and sometimes we saw solitary Bushbuck. The reedbuck showed some concern by uttering shrill, piercing whistles, and the Bushbuck barked at us like dogs. We also occasionally saw warthogs, but we did not see any Bushpigs although we knew they were there because we saw areas looking like ploughed fields where they had been foraging. On one occasion we watched some Roan Antelope swimming across a dam and, on another occasion, we saw some standing in quite deep water and plunging their heads, and horns, right under the water so they could graze on water-weeds.

The rodent fauna of the Nyika Plateau was well known, so we anticipated what species we were likely to catch, and these included some that were new to us. The most exciting of these was the Mesic Four-striped Grass-mouse (*Rhabdomys dilectus*), which is a beautiful chestnut-brown mouse with a pattern of black and beige stripes running lengthwise along its back. On the grasslands, we also caught *Mus triton* and *Lophuromys flavopunctatus*, and in one area where dense shrubs grew on the banks of a stream, we trapped just one *Praomys delectorum*. In the patches of montane forest, we only trapped *Praomys delectorum* and a species of *Grammomys*. As on Zomba Plateau, it was clear that, although the forest species very occasionally moved into shrubby areas on grassland, the true grassland species never went into the forests. We wondered if the same species, showing the same distinct habitat preferences, would be found on Mount Mulanje in the south of Malawi.

We did not expect to catch many bats on the Nyika Plateau in the cool-dry season, but we netted quite a few Dusk Pipistrelles (*Pipistrellus hesperidus*) previously only recorded, in Malawi, in another montane area north of the Nyika Plateau. We also recorded the horseshoe bats *Rhinolophus blasii* and *Rhinolophus clivosus*, in a large patch of forest known as the Juniper Forest. It was dark and very eerie working in that forest at night!

We also spent a lot of time wandering around taking photographs of the animals, and also of skulls and horns in the collection of the wildlife officers because we anticipated that we might want to make accurate illustrations of the skulls and horns one day. Then, all too soon, it was time to leave, but we had plans to camp on the Viphya Plateau (South) for a night or two, and also to spend a night as guests of Kamuzu Academy – the Eton of Africa – on our way back to Zomba.

The Viphya Plateau and Kamuzu Academy

We spent two nights camping in our tents by the beautiful Luwawa Dam on the Viphya Plateau (South). The dam is surrounded by pine plantations with some areas of grassland, some clearings and some newly built

European-style houses with ornamental gardens and groves of bananas. Like the Nyika, the Viphya Plateau did not look very African. We set lines of live-traps in a variety of habitats but did not catch any rodents of particular interest, and we set mist-nets under the trees near the dam and along its banks. The bat catch was unexpectedly interesting. We caught eight species including a second Malawian locality record for the Botswanan Long-eared Bat (*Laephotis botswanae*), which we had previously recorded in Zomba.

Moonlight, an extra special fruit bat

We also netted a female Wahlberg's Epauletted Fruit Bat (*Epomophorus wahlbergi*), which pleased us because we had received a request for the blood of one of these bats from an Australian zoologist who was working on the genetics and molecular biology of fruit bats. Australia has very strict quarantine rules and, to prevent any possibility of introducing exotic diseases into the country, any samples of blood have to be preserved in special chemicals. These preservatives had not arrived in Zomba by the time we set off north and therefore, because we could not be certain that we would catch another *Epomophorus wahlbergi* when we got back to Zomba, we decided to take this female back to Zomba with us. This meant having to feed her and keep her comfortable throughout the long, four-day journey back. We knew the Haplets would feed her, so we told them very firmly that *on no account* were they to get fond of this bat because she *had* to become a specimen as soon as the preservatives reached Zomba. We set off from the Viphya, and of course the bat ended up clinging to Lena's jumper and being given treats of banana at frequent intervals. There was never any likelihood of her flying away.

Our first stopover on our way back to Zomba was at the Kamuzu Academy, a wonderful boarding school built and funded by President Hastings Kamuzu Banda in 1981. This school is justifiably known as 'the Eton of Africa'. David had been asked to give a talk about our work in Malawi to the senior biology students, but the whole school wanted to come, so the talk was held in the school's main hall and more than two hundred students turned up. I made a last-minute decision to take the fruit bat. I would let her hang from a finger but close to my chest, and I would have my other hand near enough to restrain her if she decided to fly. I need not have worried. This amazing little animal remained totally relaxed and seemed to be fascinated by this strange new world in which she found herself. She watched everything that was going on. She accepted titbits of banana from the students who crowded around her, and she was perfectly willing for them to stroke her head. And she had only been captive for two days! How I would *love* to know what was going on in her mind.

Of course, by the time we got back to Zomba, there was no way that Moonlight (as the Haplets had named her) was going to be a specimen! We brought a small branch of a mango tree into the house, and we kept her there for a day or two before taking her outside at night to let her go. She was probably out all night, but the next morning she was back in the house! This went on for several consecutive nights until, eventually, we had to close every window so that she *had* to find another roost. We went away again soon afterwards – and so did she.

On our way back from Kamuzu Academy, we spent a night with friends in Lilongwe, and had *Chambo* and chips (of course!) in the Lilongwe Hotel. And then, back at Zomba, we had to check the Banana Pipistrelles and Little Free-tailed Bats before getting ready for our next main trip, which was to Mount Mulanje.

Venture to Mount Mulanje

We always think of Mulanje as a magic mountain! Although it should have been visible from Zomba, we did not see it during the dry season because of the smoke haze – but *everyone* told us it was there. Then, one day at the beginning of the wet season, it *was* there, floating above the Phalombe Plain as though it were something ethereal! It seemed to be floating because there were always mists around its base. It is the highest mountain in central Africa – a great massif with precipitous sides of grey igneous rock. On top, but out of sight, there is a series of basin-like plateaux at elevations of 1800 to 2200 metres above sea level, and there are a few peaks, including Sapitwa, the highest, which reaches 3002 metres. Some of the slopes, and parts of the plateaux, support stands of montane forest that have been an important source of timber in the past. One conifer, the Mulanje cedar (*Widdringtonia whytei* as it used to be called) is endemic to Mount Mulanje and it was so highly prized for its sweet-smelling timber that it is now endangered. There are no roads leading onto the plateaux, so all the timber had to be brought down by a cable (now derelict) or carried down manually as planks. The massif and all its forests are now protected as a forest reserve.

We set out from Zomba just before six o'clock, having got up at four o'clock. The Kombi was very heavily laden with all of us as well as all our notebooks, equipment and food for nine days. We went through Blantyre and then headed east across the Phalombe Plain towards the town of Thyolo. At first, Mulanje appeared as a featureless, pale blue massif silhouetted against the pale gold morning sky but, as we approached Thyolo, we started to see the water courses and deep ravines that come down the precipitous grey sides, and the dark green forests that grow around its base. It looked impossible to climb!

Mount Mulanje is high enough to influence the local weather. It is often capped with clouds and it causes rain to precipitate throughout the year although the rainfall is less during the dry season. The terrain around the base of the massif is undulating with some deep valleys, and it is ideal for growing tea. A bumpy, gravel road took us through the beautifully manicured tea estates until we came to the old forestry headquarters at Likabula that had become the headquarters of the Mulanje Mountain Forest Reserve. There, we met up with Dan and Liz Goddard and their three youngsters, who knew the mountain well and were going to come with us. There, too, we met up with seventeen porters who were going to carry all of our *katundu* (baggage) to an old forestry commission house on the Lichenya Plateau that was going to be our base. Some of the porters carried loads on their heads, some preferred their shoulders, and some carried our cases by their handles, as we would have done. David and I carried our rucksacks, quite heavily laden with cold-weather clothing and our precious, irreplaceable notebooks. None of our field records could be backed up on computers in those days!

We crossed the Likabula River and followed a path that led, at first, through miombo woodland. The porters very rapidly drew ahead of us because they were extremely fit and very used to climbing this mountain. For us, even though we were acclimatized to the altitude after our time on the Nyika Plateau, it was challenging because we knew we were facing six hours of walking from about 900 metres to about 2000 metres above sea level. The trees were wonderful: many were very old and they cast welcome, cool shadows. We passed many streams tumbling down from the plateau, and the rocks nearby were green with mosses, liverworts and tiny ferns. After two hours, we had a rendezvous and rest with the Haplets and the Goddards at 'Breakfast Creek' and, after another three hours, we met up again to have lunch. At both of these resting places, we crossed sheets of rock that were covered by raging torrents of water during the wet season and almost impossible to cross. In May, however, they were almost dry.

After lunch, we gained height and the miombo woodland gave way to montane forest and to grasslands with patches of shrubs. And then we reached the lip of the Lichenya Plateau basin and knew that the rest of the way was gently downhill! We still had a long walk in front of us and it was not long before we met our porters who had left our *katundu* at the Forestry Hut and were on their way back to Likabula. We paid them and arranged for them to return eight days later to take our *katundu* back to Likabula. *Katundu*, like *khondi*, is a lovely Chichewa word that everyone continues to use long after they leave Malawi.

It was almost dark when we reached the old hut, and we were delighted with it. It had been built in 1922 for the resident forestry officer. It was constructed of slabs of timber and had three living rooms, a bathroom containing only a tub for water, and a locked storeroom for equipment belonging to the Mountain Club. Encircling the whole hut, there was a wide *khondi* with wooden railings and steps leading down to the ground. In the main living room, there was a large stone fireplace and a fire had already been lit there by the hut's *londa* (caretaker) who also did chores for us such as washing the dishes. The Goddards were members of the Mountain Club, and therefore we were able to use the club's bushlights, cooking utensils and mattresses, so we were very comfortable. David and I slept out on the *khondi* where we could breathe the fresh air that was heavenly scented by the Mulanje cedars close by.

We always woke up to very chilly mornings and dew-drenched grasses, and we have never seen so many dew-spangled cobwebs! They spanned every path, hung from every tree and netted every tussock of grass. The mornings were almost always foggy too although the sun came through soon enough and the clouds lifted to the top of Sapitwa and then even higher.

Our first task was to set four lines of twenty-five live-traps in each of the four main habitats on the Lichenya Plateau – the montane forest and three slightly different grasslands. The forests had cedars and other trees growing up to twenty-five metres tall, and there was an understory of shrubs and creepers, or of dense forbs. We saw Blue Monkeys and Mutable Sun Squirrels in this forest, and the Haplets saw a spectacular red squirrel that we think must have been a Red Bush Squirrel, a species known to occur on Mount Mulanje. The grasslands varied. On the lower areas, three main species grew into a dense, tussocky sward about thirty centimetres high. Higher up, boulders and exfoliating rocks were common, and the vegetation was not so luxuriant. High up, there were also some grassland areas which included rock-loving plants such as the rather weird-looking shrub, *Vellozia splendens*, which has thick woody stems and only a few long, spindly leaves in the dry season. And, as on Zomba Plateau and the Nyika Plateau, there were lots of golden ever-lasting daisies.

In the trap-lines, we caught *Praomys delectorum* and *Lophuromys flavopunctatus* in the forest, and *Mus triton*, *Rhabdomys dilectus* and *Lophuromys flavopunctatus* in the grasslands. Then we established two grids, each with fifty live-traps, one in a mosaic of lower-lying grassland and forest, and the other on higher grassland with many rocks, boulders and *Vellozia* shrubs. As predicted, we found *Praomys* in the forest, *Lophuromys* in the forest and grassland, and *Mus* (but only one), *Rhabdomys* and also *Aethomys namaquensis* in the grassland. We also met some forestry workmen who had set ingenious home-made traps for rodents in the grasslands, and they were catching *Dasymys incomptus* and *Otomys angoniensis* which are trap-shy and rarely caught in the metal box-traps that we were using. We spoke to one man who told us that he caught between two and five of these rats every day during the dry season and they provided welcome protein to supplement his staple diet of maize-meal. These two species can weigh up to 218 grams and 255 grams respectively and the man told us that he usually ate about ten of these animals every week.

At night, we set mist-nets in the vicinity of the hut, but it was far too cold for bats to be flying around at that time of year.

We had very little spare time on Mount Mulanje, but we found time one afternoon to walk across the Lichenya Plateau to the precipitous edge of the mountain. With the ground falling away almost vertically, the view from the edge was spectacular! Far below us, we could see Likabula and the green tea estates and, further away, we could look right across the Phalombe Plain to Chiradzulu and other mountains near Blantyre. And even further away, we could clearly see Zomba Plateau. It was quite warm walking in the sunshine so, on the way back, when we passed a stream with a nice deep pool, we all jumped straight in. It was freezing cold but very invigorating and we warmed ourselves afterwards by basking on sun-warmed slabs of granite nearby. On another day, we took a French girl – one of a group who had joined us at the Forester's Hut – to the same viewing point, but this was a very different sort of day. It was freezing cold and a wild wind was blowing straight up the side of the massif.

It was sad leaving the Lichenya Plateau and we wondered if we would ever see it again. The porters arrived at eight-thirty to collect our *katundu*, having left Likabula at four-thirty when it was still dark. By the time the porters arrived, the Haplets had already set off, but David and I stayed to make sure nothing was left behind, and then we had a last walk to our study areas to collect vegetation with which we could create an authentic background for photographs of some of the rodents we needed to photograph back in Zomba. Then we had a leisurely walk back to Likabula with many stops to take photographs of the ever-changing zones of vegetation, and that was the end of a very special expedition.

Results of the montane studies

After completing our rodent surveys on Zomba Plateau, the Nyika Plateau and Mount Mulanje, we were able to draw the following conclusions. In Malawi, the rodents that are found above 1800 metres are not found in the miombo woodlands or any other savanna woodland habitats at lower altitudes. Consequently, the montane species are now isolated on each of the montane regions. Each of the populations of rodents on the mountains that we investigated contained six or seven species, of which four were common to all three mountains. But there were four other species that were sometimes present and sometimes absent and therefore the total species composition was different on all three mountains. The fact that the grasslands on each mountain had specialist rodents that never moved into the surrounding montane forests indicates that grasslands have existed on the mountaintops for a very long time and that they have survived there despite the climatic fluctuations associated with the comings and goings of the ice ages. Although the continuous existence of grasslands on the mountaintops has been controversial, our conclusions supported those of Professor Michael Meadows who examined fossilized pollens from the Nyika and found evidence that a mosaic of montane grassland

and montane forest had existed there for at least 12,000 years and possibly for as long as 18,000 years. The overlap in the compositions of the rodent faunas on the Nyika, Zomba and Mulanje plateaux also implies that, at some time, corridors of montane-type grasslands must have linked these regions so that some – but not all – of the rodents were able to spread from one region to another. On all three mountains that we investigated, the grasslands were patchily burnt every year and there is some evidence that this might control, to some extent, the limits of the patches of montane forest. However, the existence of the grasslands on the mountaintops for *such* a long time, and the extent to which they expanded during the cool-dry periods, suggests that their existence is not primarily due to human activities. We were very happy with the results of this project.

Last days in Malawi – and a ride on an ox-cart

After Mount Mulanje, we returned to Zomba and then had our final field-trips to Zomba Plateau, Liwonde National Park and Lengwe National Park. It was time to say goodbye to all the marked rats and mice whose lives we had been following, and to all the game scouts who had helped us so much. We had a last bat-catching session on Zomba Plateau, and then had a week of packing up and saying goodbye to all the friends we had made in Zomba.

Then it was time to get into the Kombi and head off for Lilongwe. The Kombi was heavily laden with all our record books, reference books, specimens, photos and negatives that were to come with us on the flight home, and 140 kilograms of *katundu* that we had to send as unaccompanied baggage. The latter included skulls and horns of Nyala, Impala, Southern Reedbuck and Roan Antelope, two boxes of traps (we left some at Chancellor College for others to use) and lots of souvenirs.

The Kombi, which had just been serviced, took us very willingly as far as the turn off to Liwonde National Park but then, as though it sensed we were not going in the right direction, it slowly ground to a halt! Panic set in! Would we get to Lilongwe in time to send our unaccompanied luggage off from the airport? We just managed to get to a service station and, after an hour or so of fiddling, they finally discovered that the timing was wrong because the distributor had been incorrectly rewired when the Kombi was last serviced! We only just made it to the airport in time, and there were all sorts of hassles to sort out – and then we felt obliged to drive one of the officials to his home because he had stayed late to help us through the formalities and had missed his last bus. He lived in the opposite direction to where we wanted to go, so it was dark by the time we set off along the road to Zambia, towards Namitete where we had been invited to stay for two nights with Mrs Macpherson, a wonderful old lady who we had met on Zomba Plateau. She and her son owned 'Kanongo', a large tobacco estate, and we were so glad to finally get there. I was so exhausted by then that Mrs Mac took one look at me and said, "Brandy! Immediately!" It was the first time I had ever tasted it and I am sure it saved my life!

We were woken up with cups of tea at six-thirty and had breakfast at eight o'clock followed by a walk around the garden where Mrs Mac's husband was buried, and some parts of the estate. By then, 'Kanongo' was run by Mrs Mac's son, Ewen, and it was a wonderful example of the sort of estate that was run by expatriates for the benefit of *everyone* who lived there. The estate had its own medical centre and primary school, and all the workmen were treated as tenant farmers. They and their families were well housed by Malawian standards, and they had six acres of land for their own use. However, they were expected to follow strict guidelines. Every acreage was divided into three two-acre plots and there had to be a three-year rotation from tobacco to fallow and then to a crop such as beans, tomatoes or cassava. Tobacco seedlings were supplied to the farmers, and the tobacco crop was bought by the estate owners who then processed it. The estate also had its own plantations of eucalypts for firewood, fields of maize, cattle for milk and meat, and oxen for pulling carts.

There was also a large dam on the estate and, since the Macphersons were of Scottish origin, we were not surprised that the dam had been well stocked with fish. When we told Mrs Mac that we had never tried fly-fishing, she said she would take us to the dam in the late afternoon and teach us. By mid-afternoon, we were feeling energetic, so the Haplets and I asked if we could wear shorts and run at least some of the way to the dam. We were told that this would be all right since we were on private land, and Mrs Mac said that she and David would drive down later in her very fine Peugeot. So, we started running and had almost reached the dam when we were confronted with a terrible dilemma! Coming along the gravel road towards us was a farm cart laden with children and drawn by two oxen. Did we dare do the unthinkable – in Mrs Mac's opinion – and ask for a ride? White friends of such a personage as Mrs Mac were not supposed to do such things, but we realized this might be our only chance to ride in an ox-cart and, of course, the Malawian children were wild with delight. With great haste, they tore branches off some trees and brushed out every speck of dirt (more or less) from the floor of the cart and made room for us to sit. Then they set out to show us what first-class oxen could do! They whipped them up and the oxen, which we had never seen moving at more than a very slow plod, took off at top gallop! At any moment we expected to run, at full tilt, into Mrs Mac's Peugeot coming in the opposite direction but, thank goodness, the children managed to stop the oxen just in time! The Haplets and I jumped off and immediately started running, as fast as we could, back towards the dam. Not a moment too soon! Mrs Mac and David appeared almost immediately. She must have thought we were very slow runners!

Returning to Australia

After that, all too soon we were flying home. At Sydney airport, we had to tick every box on the customs declaration except drugs and firearms! But we had permits for importing most of our specimens and the quarantine officials were extremely considerate and sensible about everything else. Then we were home again in Canberra. We looked at our diaries and realized that, throughout our entire time in Malawi, we had averaged fourteen hours of work per day on our zoological projects. This, of course, had only been possible because we could employ people to do all of the mundane chores of day-to-day living for us. We spent much of the next nine years analysing all our data and we published sixteen papers on our Malawi projects. It was somewhat difficult to smile when our laboratory manager asked us if we had had a nice holiday in Africa!

We loved working in Malawi and we are so lucky that fieldwork for us is more like a hobby than a job. As for the Haplets, before they grew up and married, they always said that their year in Malawi was the best year of their lives. We thought so too.

10

REARING BABY KANGAROOS BACK IN CANBERRA

We returned to Canberra in the middle of 1985, and the best thing about this was being reunited with Moshi who had spent the year on a farm in South Australia with friends who adored her. Just before we sent Moshi away, we had taken a tennis ball from her and hidden it in a cupboard in our bedroom. She smelled it, of course, but she could not get it. Almost as soon as she came back after her year away, she went to the cupboard and begged in front of it – her way of asking for her ball. We thought this a remarkable feat of memory – and it *was* just memory because we had put the ball somewhere else before leaving for Malawi and packed the cupboard with other things, so there could not have been any scent of the ball to remind her.

The 'kanga-babies'

Eastern Grey Kangaroos are common on farms and in bushland around Canberra. These kangaroos give birth to a single joey after a gestation of only thirty-six days. At this stage, the joey looks embryonic and weighs only about eight grams. It is naked, blind and deaf, and its hindlimbs are mere buds. However, the forelimbs are well enough developed to enable the newborn joey to climb, through its mother's fur, along a trail of saliva made by the mother, which leads from her vagina to her pouch. Once inside the pouch, the joey attaches itself to an appropriately small teat and begins to grow rapidly. It remains attached to this teat, but the teat grows and its milk changes in composition as the joey grows. Joeys begin to leave the pouch for short periods when they are about eight months old, and they leave permanently when they are about eleven months old. However, they continue to thrust their heads into the pouch and suck, from the same teat, until they are about eighteen months old. The remarkable thing about kangaroos is that females have another oestrus when their joeys are only a month or so old and, if they mate, a new embryo starts to develop but then goes into a state of delayed development that lasts until the older joey first leaves the pouch. It then develops, is born and then becomes attached to a second teat where it remains, even though the older joey spends a lot of time in the pouch for about three months more and continues to suck from its own teat for a further six months. Furthermore, about a month after the second joey is born, a third mating may occur and a third joey may remain, in a state of delayed development, until the second joey begins to leave the pouch. Therefore, at any one time, a female might have a joey at foot, another in its pouch and a third waiting to start developing *in utero*.

With this remarkable reproductive strategy in mind, it should not be surprising that stressed female kangaroos (and other macropods) throw joeys out of their pouches to lighten the load if they have to flee for their lives. Their best strategy is to save themselves because, almost immediately, they can give birth to another joey. This happened in 1989 when an Eastern Grey Kangaroo on farmland near Canberra was chased by a farm dog. Fortunately, the farmer was able to rescue the abandoned joey, and he took it to our vet friend. Then a second joey was taken to the vet after it had been removed, alive and uninjured, from the pouch of a road-killed female. Both joeys needed hand-rearing, and Lena took them on and named them Sallee and Karri. Baby kangaroos do not tolerate cows' milk, but a company in South Australia manufactured special formulae for kangaroos – a different formula for each stage of the joeys' development!

The most amusing thing was watching the kanga-babies (as we called them) learning to hop! At first, if we took them out of their artificial cloth pouches, they could barely stand on their hindlegs. Then they began to try to move by rocking onto their front paws and then swinging their huge back legs forward in front of their paws before moving their paws forward again. But, at this stage, all they wanted to do was tumble, headfirst, back into their pouches – or into anything that remotely resembled a pouch.

Before long, they started trying to hop without touching the ground with their paws, and gradually they gained speed. They were like children trying to hop around on pogo sticks. But the next problem was "How do we stop?" Well, until they discovered how to put on their brakes, they just had to crash into something! But they were undeterred. Another problem was that their hindlegs were like incredibly powerful springs, and sometimes a poor little kanga-baby would kick too hard and fly high into the air with conspicuous astonishment! You could almost hear the agonized, "Help! How do I get down again?" And then they had to learn how to turn – not as easy as you might think – and there were many falls and crashes into obstacles before they got the hang of it. But then it was absolute bliss! They just seemed to adore hopping faster and ever faster around in circles and then all around the garden.

But they were always happy to tumble back into their pouches for a well-earned nap. It seemed that any towel or rag could look like a pouch, as long as it had a few folds and creases. This happened, for example, while I was restoring an old Australian cedar desk that had been left out in the weather for several years. Its wood had to be straightened and then sandpapered down to reveal the beautiful timber under the grey, weatherworn surface, and then I applied estapol and Scandinavian teak oil and started polishing it with a pair of old flannel pyjamas. Polishing the desktop was no problem but, as soon as I started rubbing the legs, those pyjamas turned themselves into pouches and the poor kanga-babies kept trying to climb into them!

When the kanga-babies had outgrown the cloth pouches we had made for them, they moved into pillowcases that we could hang somewhere accessible. Then, when they were old enough to start leaving their pouches, we built a 2 × 2 × 2 metre dog-proof enclosure in our back garden and built a small wooden box with a corrugated-iron roof for them to sleep in. The kanga-babies were left out in the garden during the day, but they were put into the enclosure to keep them safe during the night. But the door into the enclosure was quite small – only about a square metre in size – and it opened at ground level. Consequently, it was not easy for any of us to get into the enclosure – and the kanga-babies never wanted to go inside by themselves. The only way we could entice them through the door was to go through it and then dangle a pillowcase in the doorway. This always brought the kanga-babies at top speed and then, when they approached the doorway, we would whisk the pillowcase back inside the enclosure.

151

The comical thing was that the kanga-babies would mentally record the position of the pillowcase before it was withdrawn, and then they would dive, headfirst, into the space where they had last seen it! Of course, they landed on their heads on the ground – but this never seemed to bother them. They would scramble back onto their long legs and hop inside the enclosure, and then we could close the door to keep them inside. And they adapted quite well to sleeping on a bed of straw in their box.

While raising baby kangaroos, it would be easy to conclude that they had very few brains – but they had the *biggest* hearts! They were *such* affectionate creatures! They would follow us around and loved to sit on our laps and be cradled in our arms. They got on well with Moshi and Lena's Chihuahua and, one day, when I went to visit our next-door neighbours, I came back to our garden gate to find Sallee, Moshi and Nyika the Chihuahua patiently waiting for my return. A heart-warming welcome!

Sadly, Karri had something wrong with her liver and she died. However, Sallee thrived and when she was about eighteen months old it was time to return her to the wild. Lena found a perfect solution. On the outskirts of Canberra, the old 'Gold Creek' homestead and farm had been turned into a tourist resort where people could watch sheep being shorn, sheepdogs at work and other farming activities. Near the homestead, there was a large orchard surrounded by a high fence that kangaroos could only jump if they tried very hard – but wild kangaroos that *did* manage to jump into the orchard found shelter, good grazing and some supplementary feeding and therefore they tended to remain in the orchard and they became used to people. The owners of 'Gold Creek' agreed to have Sallee, and she was released into the orchard. For a while, Lena went out there to give her some bottles of milk, and the owners also gave her some bottles and allowed the tourists to feed her as well. She thrived and became integrated into the mob in the orchard. For several months, she continued to recognize Lena and came hopping towards her but, eventually, she no longer needed bottle-feeding and became completely independent. She stayed with the orchard mob and never hopped out of the orchard, and we heard subsequently that she had had a joey of her own. It was a very satisfying ending to an animal encounter that had given many people a lot of joy.

11

STUDYING THE MARVELLOUS LIVES OF BATS IN MALAWI

Having written up the projects carried out in Malawi between 1984 and 1985, we had a lot of ideas for further research in that country and we found that our interests were starting to focus on bats. Our work on Banana Pipistrelles, together with work done by other teams in South Africa and Kenya, suggested that the bats in Kenya, which roosted in thatch, had social behaviour that was different to that of the bats in South Africa and Malawi which roosted in furled banana leaves, and we thought the differences needed further investigation. We were also puzzled by our observation in Malawi – that these bats appeared to mate in May or thereabouts and then give birth in November – because this suggested that the gestation was about six months, and this seemed far too long for a bat weighing, on average, only 3.5 grams. Therefore, we wanted to carry out a more detailed study of reproduction in this species in Malawi.

We were also inspired by a paper about wing morphology (shape), echolocation and resource partitioning in a community of bats in South Africa, and we wanted to make similar studies to see how the species of bats in various communities in Malawi manage to co-exist without excessive competition. We were also particularly keen to investigate echolocation because of the idea that every species of bat had a unique echolocation call. If this were so, species could be identified in the wild by their calls, and bat surveys could be conducted without having to catch the bats. But, first of all, the calls of every species had to be described, and it had to be proved that each species had a unique call.

David had his second sabbatical year away from the Australian National University from mid-1993 to mid-1994. We spent the year prior to this trip researching the latest techniques for recording and analysing echolocation calls and purchasing a bat detector, recording equipment and a computer that we would dedicate to this work. We also arranged for the technicians in the Zoology Department to build some of the components for a harp-trap for catching bats, and I made the other components for it. I also made a modified 'butterfly net' for catching Banana Pipistrelles, and a bucket-net for catching free-tailed bats as they dived from the roofs of houses, and we made sure that we had plenty of mist-nets in good order. We also bought specially designed, numbered bat-bands so that we could recognize each individual recaptured bat. And we needed transport and somewhere to live! After much effort and many emails and phone calls to almost everyone we knew in Malawi, David bought a second-hand VW Transporter similar to the Kombi we had had before, and he arranged for it to be in Lilongwe when we arrived. Accommodation proved much more difficult to arrange. The cost of rented accommodation in the Zomba area had skyrocketed since 1984 and was now too expensive for our limited budget. However, Sven Gruner, whom we had met briefly on the Nyika Plateau in 1985, heard of our predicament from mutual friends and granted us a reprieve. Sven had a farm called 'Kapalasa' near Namadzi, about two-thirds of the way between Zomba and Blantyre and, every year, he spent six months on the farm and six months at his home in Denmark. He very kindly said that we could stay in his house on the farm until Christmas when he and his family would be returning to Malawi. We thought this would give us time to find alternative accommodation, so we accepted his offer with much gratitude and relief. Little did we know how propitious this would turn out to be.

Eventually, David and I were ready to set off – just the two of us this time. We left Canberra at the beginning of July and, after attending the 6th International Theriological Congress in Sydney, and spending some time in Europe and England, we flew to Malawi on 21 August. After a night of fitful sleep, we woke up over the forest-savanna mosaic where the Congolian rainforest meets the Central African Savanna. It was the dry season and we could hardly see anything below us because of heat haze and smoke from the inevitable burning of the savanna. We flew to Harare but were not permitted to disembark, and then we flew to Lilongwe, a flight that took only fifty minutes. We were rather shocked by the lack of trees around Lilongwe – there appeared to have been much deforestation since we were there in 1984–85. The Malawi customs officials were as courteous and helpful as ever and, this time, all our *katundu* came off the plane very promptly. We spent a few days with friends on a tobacco estate near Lilongwe, which enabled us to do all the things we had to do before driving to Namadzi. These included trying out the bat detector. We heard several bats and were pleased that we did *not* hear any ultrasonic sounds made by all the crickets and other insects that were filling the air with noise. A bat detector lowers the frequencies of ultrasonic sounds so that they become audible to humans, and we wondered if we would ever be able to recognize the calls of any species just by listening to them through the bat detector. Some people claim they can do this.

The next day, a Monday, was dedicated to all the administrative chores that had to be done in Lilongwe and, the day after that, we loaded the Transporter and set off for Namadzi and 'Kapalasa' Farm. The Transporter went well except for being slow on the hills. The road across the Central African Plateau was busy with people walking, cycling or selling things, and we saw many children playing with home-made cars and trucks made of wire, or using sticks to roll hoops over the ground. The refugees from Mozambique were gone, and the houses that used to shelter them had been vandalized and everything useful had been removed. At Ntcheu, we started descending the escarpment and we had wonderful views across the Upper Shire Valley. Then we crossed the Shire River and smiled as we saw 'Happold's Hill' in Liwonde National Park on our left. And soon we were climbing the escarpment leading up to the Shire Highlands and we were pleased that these hills were still covered by dense miombo woodland. Soon after, we were within sight of Zomba Plateau and then we reached Zomba. We drove to Chancellor College, met a friend in the Biology Department, collected some mail and then set off for 'Kapalasa' Farm.

It took forty-five minutes to reach the farm – longer than we had expected – and the last five kilometres was along a narrow dusty road that looked likely to be horribly muddy in the wet season. On either side, there were newly ploughed fields. We passed a cluster of houses where the 'Kapalasa' farmworkers lived, and then we drove into the garden around Sven Gruner's house. Everything looked very dry and most of the plants seemed to be aloes although there were some bare deciduous trees and a few palms. The house was a rambling, converted tobacco barn but it was very comfortable. Part of our arrangement with Sven was that we would employ his housekeeper and the staff who looked after the house and garden and the Gruners' five dogs. We met Joyce Malombe who was the cook-*cum*-housekeeper and it appeared that she did not speak much English – but this was just because she was very shy when we first arrived. She did not do any shopping, so there was little food in the house, and nothing was prepared for our arrival. The phone did not work and there was no gas for the stove. We went to bed feeling very isolated and rather depressed.

Next day, we drove into Namadzi to arrange for the telephone to be connected to 'Kapalasa', and then into Zomba. At the bank, we discovered that our salary had not been sent from the Commonwealth Trading Bank, nor had our field expenses been sent, so we had only K8.50 in the bank! The exchange rate was not good – Aus $1 = K3 (three kwatcha). Luckily, the manager let us have an overdraft, so we were able to go to the Kandodo and PTC supermarkets to stock up on food. Then we went to the market where we bought a wicker basket and vegetables, and then to Chancellor College where we obtained some work essentials such as 70 per cent alcohol and Bouin's solution (a special preservative).

That evening, we cooked our supper on the Trangia stove and had yoghurt and banana for dessert. We tried out the bat detector and heard nothing at all and there were no insects at the windows. It was very disconcerting.

The following day was very busy. We went to Chancellor College to collect more equipment, mail and a fax from the departmental secretary at the Australian National University, which told us that the Commonwealth Trading Bank had, "overlooked our arrangements for funds"!! Then we went to the Forestry Research Institute of Malawi to obtain permits for working in forestry reserves. In Zomba, we bought yellow cloth for bat-bags and arranged for a *khondi* tailor to make them for us, and we went to the sawmill on the road to the plateau and arranged for them to make fourteen dowel poles, each 2.5 metres long, for use with the mist-nets. We also went to the telecommunications office and were assured that the phone was now reconnected to 'Kapalasa'. It rained in Zomba that afternoon, which was very unusual for the time of year, and everyone was wearing thick coats and carrying brightly coloured umbrellas.

When we reached 'home', we tested the phone and it was *not* working, so we walked along the road to 'Top House', the second house on 'Kapalasa', and introduced ourselves to Murray and Alison Pedder who were renting it. They were a young couple from South Africa. The Pedders were very friendly and we liked them immediately. They said we could use their phone to arrange for ours to be properly connected. That night, we lit a fire in the fireplace at 'Kapalasa' and sat around it. David had a Carlsberg 'Green' beer, and we listened to one of the Prom concerts on the BBC. Tomorrow, after all the delays and frustrations of getting organized, we were going to start our first bat project!

The Banana Pipistrelle project on 'Kapalasa' Farm

Our first impressions of 'Kapalasa' as a study area were very negative, but the farm had one great asset that we started to exploit immediately – a plantation of dwarf banana plants that might have been planted just for us! It comprised a grid of 138 clumps of banana plants laid out in seventeen parallel rows with eight to ten clumps in each row. Because the plants were a dwarf variety, it was possible to completely enclose a furled leaf in our modified butterfly net, pull a string to tighten the base of the net around the base of the leaf, and then remove every bat in the leaf. It was also possible to put a net over *every* furled leaf, so the entire plantation could be censused. The 'Kapalasa' plantation had another advantage – some areas were better shaded by other trees, better watered and had more fertile soil, and these areas grew banana plants that were healthier and taller and produced more furled leaves. In other words, the reliability of the clumps (as providers of furled leaves for the bats) showed considerable variation.

We conducted a census every two weeks except from mid-November to the end of December when the females were giving birth and suckling very young babies. We worked up and down the rows, collecting all the roosting bats and keeping the occupants of each leaf together in one of the bat-bags. Every clump was given a number, and we recorded the number of furled leaves suitable for occupation that were produced by each clump, and we used this as an indicator of the clump's reliability. At the end of the census, the bags of bats were taken back to the house and then every new bat was banded with a numbered band, and its weight, sex and reproductive condition were recorded. We also recorded which banana clump each bat was roosting in, and which bats it was roosting with.

The study yielded a huge amount of data – far too much to record here. It confirmed that females that were about to give birth or suckling their young roosted singly or with one or two other females and their young, and only very rarely roosted with a male. During this time, the males almost always roosted singly. At all other times, the males either roosted singly, or roosted with one to ten females (and rarely another male). We ascertained that the formation of groups was not due to a shortage of furled leaves. On most occasions, each male was found roosting with different females and each female with different males. This indicated that group-membership was highly labile and that the mating system was promiscuous – potentially every male could mate with many females and every female could mate with many males. Because furled leaves open after one to three days, strict roost-fidelity is not possible for Banana Pipistrelles, but we found that most of the males showed fidelity to a small number (less than five) of reliable clumps that were close together and had one or more furled leaves available at all times. The males that showed the most roost-fidelity were those that roosted in the most reliable clumps. In contrast, the females did not show roost-fidelity, but the most reliable clumps attracted more females than would be expected if they chose clumps on a random basis. Another interesting observation was that when the males became sexually active, some males attracted – in total – more females than other males did although the attractiveness of individual males varied from census to census. This made us wonder if the most attractive males were those who were roosting in the most reliable clumps – and was this because the females were attracted to those males, or were they attracted to the most reliable clumps? It turned out that, although females apparently prefer to roost in the more reliable clumps, the males that roosted in the more reliable clumps did not necessarily attract the most females. This implied that females were attracted to particular males because they had attractions other than their roosts. Possibly the females were attracted to the most vocal males as was the case with the thatch-roosting bats that were studied in Kenya, but we were not able to study this.

We were interested to compare the behaviour of the leaf-roosting Banana Pipistrelles with that of *Myotis bocagii*, a bat that also roosts in furled banana leaves, and we also wanted to compare the leaf-roosting Banana

Pipistrelles with those that roosted in holes in thatch. In Gabon, André Brosset found out that adult female *Myotis bocagii* greatly outnumbered adult males and they roosted in groups with stable female membership that also included one adult male. Such groups are called harems. The duration of a male's tenure of a harem could sometimes be as long as a whole year.

In Kenya, Tom O'Shea found that bats that were thought at the time to be Banana Pipistrelles (*Pipistrellus nanus*) roosted in holes in thatch. As in the leaf-roosting bats, the males roosted singly, or with groups of one or more females that were highly labile and therefore not harems. However, there are some substantial differences in the roosts and the behaviour of the leaf-roosting and thatch-roosting bats. The main differences are that, for the thatch-roosting bats but not the leaf-roosting bats, (1) some of the roosts are too small to accommodate females as well as a male, (2) spacious roosts are limited in number, (3) successful competition between males for spacious roosts confers the advantage of access to females during the day and (4) successful defence of a spacious roost by a male has the potential to demonstrate that he is fitter and has more desirable genes. O'Shea also discovered that the males vocalized at their roosts, and that females were attracted to the most vocal males. Furthermore, he observed wounds on the males that indicated that they sometimes fought while competing for the most spacious roosts. These observations led O'Shea to conclude that the mating system of the thatch-roosting bats resembled resource-defence polygyny. O'Shea was not convinced that the leaf-roosting Banana Pipistrelles and his thatch-roosting bats were the same species and later, more evidence showed that he was right: his bats are now referred to as Samburu Pipistrelles (*Pipistrellus* cf. *helios*).

As well as following the reproductive condition of the bats we censused in the plantation, we collected Banana Pipistrelles from other localities in the Namadzi area, and dissected these to obtain more detailed information, and then the reproductive tracts were sent to a colleague in South Africa who prepared slides and examined the tracts under a microscope. This study showed that spermatogenesis began in the hot wet season (February to April) and that mating occurred between mid-June and early July in the middle of the cool dry season. The interesting thing was that sperm were then stored in the uteri of the females until these females ovulated in August. Therefore, although the interval between mating and giving birth was about twenty-three weeks, the period of developmental gestation was only about ten weeks. This reproductive chronology was very advantageous because it meant that each of the two most energy-demanding events – spermatogenesis and lactation – could take place in the wet season when insects were most available for food. This was the first time that sperm storage, as a means of lengthening the interval between mating and parturition, had been recorded in an African bat. It was also very interesting that, because the males and females roosted together but with different individuals on different occasions during the period between mating and ovulation, it was possible that multiple matings with different partners took place and that there was competition between the sperm from the different males.

We ended up being very frustrated by the results of these two studies. We did not know if some of the males in the 'Kapalasa' plantation sired more offspring than others, and therefore we could not correlate such things as roosting in the most productive clumps and attracting the most females with the reproductive success of the males. Furthermore, we did not know if siblings sometimes had different fathers, which would have proved the occurrence of multiple matings and the likelihood of sperm competition. When we came back to Australia, we learned that, if we had taken minute samples of wing-membrane from all of the bats in the plantation, we could have used DNA fingerprinting to answer all these questions! As far as we know, no one has as yet repeated our observations and done the DNA work to answer these questions, and this is a great shame.

Resource partitioning in communities of bats

Most of our time in Malawi in 1993–94 was devoted to the collection and analysis of data about those aspects of bat biology that enable many different species of microbats to live together in communities without excessive competition for the available resources. This study was multifaceted. We wanted to carry out more surveys of the bat communities on Zomba Plateau, Zomba, and Liwonde National Park and we wanted to survey the bats on 'Kapalasa' Farm. We also wanted to conduct a short-term survey at Chiromo in the Lower Shire Valley (because the bats at that locality were already fairly well known). In addition, to obtain more information about Malawi's bats, we wanted to conduct short-term surveys at Ntchisi Forest Reserve, Nkhotakota Game Reserve, Thyolo, and several localities near

'Kapalasa' Farm on the Shire Highlands. For all of the species found at all of these localities, we wanted to record and describe their echolocation calls, wing morphology and flying abilities, diets and foraging behaviour, and domiciles (where they roosted during the day).

Collecting things has always been a passion, and it was the same with collecting data. One can never get enough of it. This year in Malawi was particularly wonderful because each night of 'batting' and every locality we visited, had the potential to yield new locality records, new echolocation calls, new wing tracings and so on. And, as our collections of data grew, so too did our understanding of these fascinating creatures called bats, and that was the greatest reward. I hope I can share this excitement with you.

Echolocation

Unlike fruit bats, microbats use echolocation to perceive their surroundings and hunt their prey that is mostly insects. African microbats are currently placed in ten families, of which eight have widespread distributions in Malawi and were studied by us. These families, and the number of species in those families that we worked on, are the horseshoe bats (Rhinolophidae: seven species), Old World leaf-nosed bats (Hipposideridae: three species), trident bats (Rhinonycteridae, one species), sheath-tailed bats (Emballonuridae: one species), slit-faced bats (Nycteridae: four species), free-tailed bats (Molossidae: two species), vesper bats (Vespertilionidae: nineteen species) and long-fingered bats (family Miniopteridae: uncertain). To study the echolocation calls of these bats, we used a bat detector that makes the calls audible to humans. We could simply switch the bat detector on and listen to any bats that were present, or we could connect the bat detector to a tape-recorder and record the calls. Then we could add a spoken comment after each recording, giving information such as the species of the bat and where it was and what it was doing when the recording was made. Another piece of equipment enabled us to connect the tape-recorder to a computer to produce a sonogram of the echolocation call – this being a graph of frequency against time. Sonograms, in effect, enable you to *see* what a sound *looks* like, and then the 'shape' of the call, its frequencies, its duration, and the time between consecutive calls can be measured.

Most species of bats in Malawi use echolocation calls that start at very high frequencies and end at much lower frequencies a few milliseconds later. These are called frequency modulated or FM calls. Many vesper bats echolocate with FM calls whose sonograms look like hockey sticks, but there are many other variations in FM call shapes. In contrast, the echolocation calls of horseshoe bats, leaf-nosed bats and trident bats are of longer duration and most of the call is pitched at a constant frequency (CF calls). Sheath-tailed bats make echolocation calls that are almost constant in frequency, and free-tailed bats make calls that have both FM and CF components, depending on what the bat is doing at the time. Slit-faced bats make very quiet calls that are quite different in their characteristics, and these calls could not be analysed by our equipment.

The different sorts of echolocation calls have different uses. CF calls can be used to perceive insects that are fluttering their wings. When echolocation calls are reflected back from fluttering wings, the echo changes in frequency (pitch) as the wings move towards and then away from the listening bat. Exactly the same thing happens when a police car comes towards us and then moves away from us with its siren blaring – the pitch of the sound changes. This is called Doppler shift, and bats that exploit this make long CF calls whose echoes from fluttering wings oscillate in frequency rather like a sustained trill on a musical instrument. These bats specialize in hunting for fluttering moths and other insects whose wings are fluttering. In contrast, FM calls enable bats to perceive the size, shape and texture of potential prey, and how far away it is. Bats making FM calls can distinguish texture so precisely that they can detect a stationary moth resting on the bark of a tree! Many species use calls that have both FM and CF components, and some species can change from one sort of call to another depending on what sort of information is required. It would be wonderful to know how bats 'visualize' things in their brains. I think that, in some ways, echolocation by bats is like humans using a torch in the dark. We can negotiate obstacles in a dark room by flashing a torch on and off, but we can improve our perception by flashing the torch more often. Bats exploit this principle by emitting echolocation calls at a greater rate when they are closing in on their prey and need more information, and this results in a 'feeding buzz' that is easily recognized by humans using bat detectors.

During this year in Malawi, we recorded the echolocation calls of thirty-one species, and we used echolocation calls to distinguish one species, *Rhinolophus swinnyi*, from *Rhinolophus simulator*, which is almost identical in its anatomical characters.

Wing morphology and flight characteristics

The wings of bats are as variable in shape as the wings of birds, and their flying abilities, and where they fly, are just as varied. We made tracings of the outlines of the flight-membranes and bodies of all species of bats that we encountered. Then, from the wing-areas and wingspans, we could calculate the aspect ratios of the wings – this being a measure of their length relative to their width. Species with long, narrow wings have high aspect ratios and vice versa. We also weighed every bat, which enabled us to calculate its wing-loading (the mass of the bat divided by the area of its wings). Aspect ratio and wing-loading influence flight characteristics in the following ways. High speed is correlated with high wing-loading, and bats (such as free-tailed bats e.g. *Tadarida*) with high wing-loadings can only fly at high speeds, which explains why they must dive from high roosts in order to gain sufficient speed for flight. Aspect ratio is related to the energetic cost of flight, the cost being lowest when the aspect ratio is highest. This means that free-tailed bats, which have high aspect ratios, can fly and forage for long periods of time without having to land and rest. The ability of a bat to turn and the radius of its turn, depend on its speed – the radius of turn being smallest when the speed is slowest. For this reason, some bats swoop upwards to lose speed before turning, and some almost stall and therefore achieve 'stall-and-twist' turns that are extremely tight and wonderfully acrobatic to watch. The rate at which a turn can be initiated is referred to as agility, and agility increases with fast flight and high wing-loading. The minimum space required for a turn at a given speed is referred to as manoeuvrability and, because the radius of a turn increases with increased speed, manoeuvrability is favoured by slow flight and therefore low wing-loading. Manoeuvrability in confined spaces (e.g. in places that are cluttered by vegetation) is also favoured by short wings and small overall size. These are just some of the characteristics of flight that are determined by aspect ratio, wing-loading and overall size.

We wanted to make quantitative descriptions of the flying abilities of the different species of bats, so we conducted one-minute-long flight-tests in a 1 × 1 × 1 metre cage made of a light wooden frame and mosquito netting. We recorded how many times a bat circled the cage during each flight-test, and whether or not it could land on the floor and take off again.

We also recorded whether or not it could fly a figure-of-eight path within the cage – only bats whose radius of turn was twenty-five centimetres or less could do this. This proved very satisfactory and it clearly separated the flying abilities of the bats into vastly different categories. For example, the free-tailed bats, *Tadarida pumila* and *T. condylura*, which have high aspect ratios and medium or high to very high wing-loadings, were unable to fly in the cage other than from one side to the opposite side. Neither could the sheath-tailed bat, *Taphozous mauritianus*, which has medium aspect ratio and medium wing-loading. At the other extreme, the slit-faced bat, *Nycteris hispida*, which is a small bat with very low aspect ratio and very low wing-loading, could fly at least ninety-three times around the cage, and could easily fly figure-of-eight paths and take off from the ground. It was fantastic watching this species in the cage!

Foraging strategies – and the bat who created havoc and a legend!

Finding out where microbats forage is very difficult because they forage at night or at dusk when it is difficult, if not impossible, to see them. However, some other bat biologists had observed foraging behaviour by lightly gluing a tiny pill-capsule filled with fluorescing chemicals to the belly fur of a bat. This is called light-tagging, and we were keen to try it.

The first species we light-tagged was *Myotis bocagii*. We released the light-tagged bats where we had netted them on the bank of a large dam. It was amazing! We watched these bats foraging very close to the surface of the dam so that we saw the reflection of the light as well as the light itself. Very frequently, the two lights seemed to touch as a bat flew down to the surface of the water. We recorded and listened to the echolocation calls of the light-tagged bats, so we could tell when they were searching for prey or homing in to attack it, so we soon realized that these bats were using their longish hindfeet to gaff insects that were floating on the surface.

It was fascinating watching slit-faced bats such as *Nycteris hispida* and *Nycteris macrotis* flying very slowly, and with great manoeuvrability, close to the ground and in and out of tall grass stems and the branches of low shrubs. The foraging flights of these bats were very short and then the bats landed and swivelled around presumably surveying their surroundings for prey. We watched one *Nycteris hispida* for thirty-eight minutes and, during this time, it spent only four and a half minutes in flight.

It was also fascinating watching the free-tailed bats, *Tadarida condylura* and *Tadarida pumila*, while they were foraging in open spaces high above the ground, but one of these light-tagging sessions created unanticipated havoc that is *still* talked about in Malawi! On this occasion, we had four *Tadarida condylura* that had been caught as they emerged from under the roof of one of the houses in the farmworkers' village on 'Kapalasa'. We took them into a large field on 'Kapalasa' at about eight o'clock by which time it was pitch dark and all the farmworkers and their families had gone home. We light-tagged the bats, one at a time, threw them into the air so they could dive to get enough speed to start flying, and then watched as they circled to gain height and then began foraging. We were making recordings of their echolocation calls so we could hear as well as see when each bat located an insect, approached it and then attacked it. These bats emit differently 'shaped' calls during each of these activities. The bats foraged high overhead either in straight lines or very wide circles, and it was awe-inspiring watching how fast they flew and how they could suddenly swerve or dive to pursue an insect. We always tried to glue the light-tags onto very few hairs so they could be groomed off fairly easily, and the first three bats that we watched dislodged their tags in less than a minute. The fourth bat, however, kept its light-tag on and we watched it for just over a minute until it flew off, in a straight line, towards the compound where it roosted. Then we went home to bed and it was not until the next morning that we heard what had happened next. At nine o'clock, that bat – still carrying its light-tag – started foraging in wide circles high above the tobacco barns on another estate and the Malawians who were on the night shift stoking the fires where the tobacco was being cured saw it and panicked! They could not tell how big this light in the sky was, or how far away it was, and they feared the worst. A group of them jumped on to a lorry and drove away at high speed to find the owner of the barns. They shouted, "*Bwana! Bwana*! Come quick. Something is flying in the sky and the men are going!" The *bwana* (master) leapt onto the lorry and rushed back to the tobacco barns. He found some of the men praying on their prayer mats, others were lighting a ring of fire around the barns, and some were running away as fast as their legs could carry them! The *bwana* looked up and, for some minutes, watched the bright light circling high above the barns. And then the bat dislodged the capsule and it plummeted to the ground close to where he was standing! There it was, shining brightly in the grass! The *bwana* was ever so brave – he picked up a large stick and tried to wallop the capsule! He missed and tried again. And missed again! Then he started to approach it very warily and, when it did not move, he put out a finger and touched it. He did not get burnt or bitten, so he gingerly picked it up and then recognized it as a pill-capsule. It occurred to him then that perhaps those crazy people staying at 'Kapalasa' had put the light onto a bat, but he was not able to confirm this, and he must have told many people about this mystery because, years later, we heard that he was still known in Malawi as Mr UFO!

Of course, when we heard what had happened, we were appalled! Everyone on 'Kapalasa' and the neighbouring farms had been so friendly and kind to us that the last thing we wanted to do was to terrify them. So, I went to the Malawian estate manager of 'Kapalasa' and confessed. However, by the time I finished the story, he was laughing so much he could hardly stop and I was much relieved. I realized that if he could laugh, so too could all the other Malawians. So I asked him to tell all the 'Kapalasa' people what we were doing and what had happened, and we asked him to tell everyone on the neighbouring estates as well. We were very grateful that this story had a happy ending.

To make sure that the people at the 'Kapalasa' farmworkers' village knew about the light-tagging, we arranged a demonstration. We caught five *Scotoecus hirundo* as they emerged from under the roof of one of the houses and, the next night, we took them to a clearing about 800 metres away where we light-tagged them and released them. We hoped that they would return to the compound, and all the children were allowed to stay up to see if they did. The light-tagged bats foraged nearby for a while, and then one of the bats *did* return to the compound and we heard the children shrieking with excitement and delight. They told us that the bat had flown around between the houses and sometimes over them, and they had watched it for nearly two minutes before it flew out of sight.

All in all, we light-tagged sixteen bats representing six families, and we also observed and recorded other bats if they foraged after being released while it was still light enough to see them. It was magical watching the lights of tagged bats dancing through the bush or sailing across the sky overhead. Sometimes there were so many fireflies that it was hard to keep track of the lights on the bats! We listened to the bat detector and tape-recorded a running commentary on what was going on. It was exciting – especially when we were working in national parks where we also had to watch out for lions!

We recognized eleven different foraging zones exploited by bats. These were as varied as (1) open areas high above the ground and trees, (2) over open water within a metre of the surface, (3) in moderately uncluttered places under the canopies and between the trunks of trees, and (4) in densely cluttered places within the foliage of trees, or within understory vegetation in forests or woodlands. We also recognized seven different foraging strategies. These were as varied as (1) fast-hawking in open spaces where flying insects are detected at long range, pursued at high speed with great agility, and eaten on the wing, (2) gleaning where prey is plucked from foliage, tree trunks, walls, rock faces or the ground, (3) trawling where the bat's hindfeet are used as gaffs to take prey from water surfaces and (4) chasing, which is the pursuit of non-flying prey such as cockroaches by bats that are scuttling over the ground or over horizontal or sloping surfaces such as tree trunks or floors and beams in buildings. Most bats specialize in just one foraging strategy and forage in just one zone.

We predicted that, if the species of bats in a particular community had different sizes and weights, different echolocation calls, differently shaped wings and consequently different flying abilities, different diets and different foraging strategies, and different domiciles, they could co-exist without much competition between them. We studied the communities of microbats found at five localities: Zomba Plateau (1520–1800 metres: ten species), 'Kapalasa' Farm (1000 metres: nineteen species), Zomba (800–900 metres: twenty species), Liwonde National Park (500 metres: twenty-seven species) and Chiromo (less than 100 metres: twenty-eight species) and found that this was indeed so.

'Kapalasa' Farm and the conservation of mammals

Sven Gruner, who owned and managed 'Kapalasa' Farm, was a wonderful wise man who believed that it was important to conserve the plants and animals on his farm as well as running it economically. He also found that this made the farm a more attractive and interesting place to live in.

'Kapalasa' Farm lies on the Shire Highlands and, prior to the 1800s, the highlands were covered by miombo woodland with riverine forests bordering the streams. Since 1900, the density of humans has increased dramatically and most of the woodlands have been converted to subsistence farms or large commercial estates, and natural vegetation only occurs in a few forest reservations and in remnants on some of the farms. 'Kapalasa' began as a tung estate but, since it was bought by Sven Gruner, it has been used for growing tobacco and seed-corn. Sven imposed some wise rules – no vegetation was to be cut down along the streams and on the dambos (marshes), no living *or dead* trees (except eucalypts) were to be cut down for firewood without special permission, and the vegetation on termite mounds and on the rocky hills and ridges was to be left untouched. Hunting, except for rodents, was not allowed. Arable areas on the farm were divided into fields on which crops and fallow were rotated, and every field was crossed by one or more wide grassy bunds that prevented soil erosion and gave access to the fields. The bunds also enabled ground-dwelling animals to move from field to field and between the areas of natural grassland, and also between the woodlands and forests that grew beside the streams, on the termite mounds and on the rocky hillsides.

'Kapalasa' had a wide variety of different places where bats could forage, and it had hollow trees, trees with foliage and loose bark, palms, banana leaves, caves, rock crevices, bridges, culverts and different sorts of buildings in which they could roost during the day. There was also a small dam on the stream that ran through 'Kapalasa' and a large dam on one of the adjacent estates. We recorded nineteen species of microbats and three species of fruit bats on the farm and, of these, five species are 'rarely-recorded' in Malawi, one is 'seldom-recorded', and five are 'uncommonly-recorded' (see Happold & Happold 1997, for definitions).

We also set live-traps for rodents and shrews in all of the different habitats on 'Kapalasa' and we made a list of any other mammals that were known to occur on the farm. The total was impressive. There was one species of shrew, one hedgehog, one sengi (or elephant-shrew), one monkey, one hare, thirteen rodents, twenty-two bats, three carnivores and one species of duiker (a small antelope). Of the species of small mammals that were live-trapped, nine were rodents and one was a shrew. The number of species trapped on the termitaria varied from one to three, whereas the larger remnants of forest had up to five species. Also, the number of individuals in the forest remnants increased with increased area of the remnants, which showed that the conservation value of farms is enhanced by putting aside some larger areas of natural vegetation.

So, 'Kapalasa' Farm conserves at least twenty-two species of bats and seventeen species of rodents, shrews, hares and sengis (total thirty-nine).

For comparison, Liwonde National Park, which is the nearest national park to 'Kapalasa' (although it is only 500 metres above sea level in the Shire Valley) is known to have twenty-seven species of bats and fifteen other small mammals (total forty-two). However, there are no formally protected areas where mammals are protected on the Shire Highlands and, furthermore, 'Kapalasa' has six species of bats and six species of rodents that have not been recorded in Liwonde National Park. Because 'Kapalasa' conserves some species that are not protected elsewhere, and because it conserves some species that are either 'rarely-recorded' or 'seldom-recorded' elsewhere in Malawi, it plays an important role in conserving and maintaining the diversity of mammals in Malawi. We felt very privileged to know Sven Gruner who set such a wonderful example to other estate owners in Malawi and elsewhere.

Our year on 'Kapalasa' Farm

Although we were rather depressed when we first found ourselves living on 'Kapalasa', we soon changed our minds. When we got to know Sven's staff, and they got to know us, we began to respect each other and then we liked each other very much. Joyce and I became good friends and one of the nicest things about this friendship was that we could be discussing and planning household affairs very formally one minute, and then sharing jokes or teasing each other the next.

We found our time on 'Kapalasa' very interesting but working in Africa can be very challenging and our first six months were full of problems. Originally, we thought we could only stay in Sven Gruner's house until just before Christmas, but we could not find any other affordable accommodation and this was a nightmare. Fortunately, we were able to spend the Christmas period at 'Top House' while Murray and Alison Pedder were away, but, during a period of heavy rain, its roof leaked and the sodden ceiling over one room fell in! Also, the power and the telephone kept going off – and the Transporter gave us endless trouble until a cousin of David's in Germany sent out a new clutch plate and some other essential spare parts. Because we had so much electronic equipment, we found that we urgently needed another power board with *Australian* sockets of course, so we sent a fax to our laboratory manager at the Australian National University and his reply was, "Why don't you buy one in Malawi?" One of our tape-recorders failed and we had to get it replaced from Australia. Several components of other crucial equipment were faulty and they also had to be replaced! A film sent overseas for processing was lost in the mail. Oh dear! During those first months, we seriously thought that we would have to abandon everything and return, defeated, to Canberra.

But there were some delights to console us. In November, Jonathan came to stay with us for two months and then Lena decided she could not bear to miss out on the joys of Malawi, and she came out for the last five weeks of Jonathan's stay. One or both of them came with us on field-trips or they visited friends or just enjoyed living on 'Kapalasa'.

Also, we did a lot of interesting 'batting'. There was a large dam about a kilometre away on the neighbouring 'Kapino' estate and we netted ten species there. Another place we visited quite often was the compound where most of the 'Kapalasa' farm workmen lived with their wives and children. We set mist-nets to catch bats as they emerged from under the roofs of the houses, and the children were always vastly entertained by this. I spent a lot of time showing them the bats and before long, like most people in Malawi, I was given a nickname – *Mama Mleme*. Mother Bat! I loved it! It was so nice hearing the excited voices of the children shouting, "*Mama Mleme abwera! Mama Mleme abwera!*" Mother Bat is coming! We never found out what they called David!

One day, while we were staying at 'Top House', we gave a talk about bats in Blantyre. We decided that David would talk first and, after thirty minutes, I would take over. David began with a quick summary of the subjects we intended covering, and then said that our main objective was to make everyone realize that bats were really useful, nice, warm, friendly, cute and very likeable little animals. At that stage, I found myself surrounded by an extremely sceptical audience who were certain that they would not be swayed by anything we could say about bats! So, when I finished my half of our talk, I told this audience that we had tried to make them like bats, but we knew we couldn't succeed just by talking and showing slides. But I added that we were going to convince them by introducing them to two fruit bats (*Epomophorus labiatus*) and a vesper bat (*Scotophilus dinganii*) that we had captured only the night before. I pointed to a corner where we had the bats in cloth bags. The response was amusing but predictable. The vast majority of the audience made a beeline for the opposite corner where coffee and tea were being served. Only a few were curious enough to have a look at the bats – but the bats were wonderful! They took fruit or termites from the people and let themselves be stroked. The *Scotophilus* was even happy to be held in the hands of some of the audience. It was interesting listening to comments such as:

"I really didn't think that bats looked like that!"

"It's more furry than I expected!"

"Oh! It's so warm and soft!"

"It's actually got a really nice face – a bit like a puppy's!"

"Will it really take food from my fingers?"

And then various members of this group went off to the opposite corner and dragged reluctant friends to see the bats – and their reactions were the same – incredulous but delighted!

Sadly, on Boxing Day, Lena and Jonathan had to return to Australia. We felt very flat for a while but, thank goodness, our fortunes changed with the coming of the New Year! The best thing was that Sven Gruner said that we could come back to stay in his house when it was time for us to leave 'Top House'! This was an enormous relief! It meant that we could continue the fortnightly censusing of the Banana Pipistrelles, and we could continue the 'batting' and small mammal trapping on the farm, which was proving to be more and more interesting. But we had a number of field-trips to other places of interest.

Our field-trips while we were based at 'Kapalasa' were sometimes back to our old haunts, but we also went to several localities that were new to us. These included Ntchisi Forest Reserve, Nkhotakota Game Reserve, Thyolo and Chiromo. Visits to our old haunts were very nostalgic and it was interesting to observe the changes that had occurred since 1984–85. And going to the new localities was very exciting, but what thrilled us most was adding to the list of species of bats and small mammals that we were able to find, and gradually accumulating the data we needed for our studies.

Returning to old haunts

Of course, we went back to Zomba several times. Our aim was to visit localities where we had caught species of bats in 1984–85 that we now needed for our new projects. These included the caves on Chinamwali Hill and so we went there with some helpers who lived nearby. The hillside was almost unrecognisable! Most of the trees had gone and much of the hillside was terraced with subsistence gardens of maize, cassava, tomatoes, chickpeas, beans and pumpkins. Eventually we found one cave that was occupied by 60–80 horseshoe bats, but they were hanging from a high ceiling about ten to fifteen metres above our heads. We made a lot of tape-recordings as they were disturbed and started flying, and one of the boys climbed high and caught four of the bats in our butterfly net. They were all *Rhinolophus fumigatus*, which was very useful. The next day, we set mist-nets around the garden in central Zomba where we were staying, and also in a plantation of old trees near the golf course, and along the hedge-like shrubs and creepers at the boundary of the golf course. We caught thirty-four bats! They included *Rhinolophus blasii*, which we recorded and light-tagged, and it was interesting to see that they foraged mainly along the 'hedge' at the edge of the golf course.

We also had a nostalgic visit to Malonje village near Zomba, where I had obtained most of the Banana Pipistrelles we had studied in 1984–85. There we met Elwin Mbanga who had been one of our 'bat-boys' at that time, and it was great to see him again. We enlisted his help for old times' sake, and we obtained a useful sample of ten Banana Pipistrelles. We were given a very warm welcome in the village by people who remembered me coming there to catch bats in 1984–85, and then Elwin took us to meet his great uncle who had a house nearby. We asked if we could take a photograph of the old man and he agreed provided we would wait until he had changed into his best clothes. He brought a mat from his house and sat down on it, but then he got to his feet and asked us to wait because he had remembered something he needed from his house. He came back with his hoe and placed it beside him when he sat down again on his mat. His words were unforgettable. "This," he said, pointing to his hoe, "is so that you will never forget that I am a farmer!" We promised to send him copies of the photos we took, but sadly the old man suddenly died before he could see them. He was a wonderful person – contented with his simple life, proud of his work, dignified and very warm-hearted. A man one deeply respected.

During the year, we also had three 'batting' trips to Zomba Plateau. In mid-January, we spent a week staying in a cabin at 'The Stables' – the home of Brian and Jayn Burgess who had been friends since 1984–85. On our first night, we set mist-nets around the edge of Mulunguzi Dam and in the forest around the Mulunguzi Stream that runs into the dam. We netted sixteen *Pipistrellus rueppellii* that night, which was wonderful because we were able to get all the wing tracings we needed for this species as well as echolocation calls and light-tagging data. These are very beautiful bats with silky grey dorsal fur, pure white ventral fur except for the throat, which is sometimes pale rusty-brown, and black wings. The next night, we set mist-nets around the Trout Hatchery but only caught three bats but, on the third night, we went to a locality that was new to us – the Chagwa Dam, which lies in a deep valley surrounded by montane evergreen forest and pine plantations. It is a very beautiful dam and it was so quiet and peaceful there. The reflections of the dark green trees in the still waters of the dam were magical. We set four mist-nets, three close to the dam and one on the track leading down to the dam, and it was very successful. We netted twenty bats representing six species, and we were able to get many recordings of the echolocation calls of these species as they flew towards the nets in which they were caught. This was very useful because the identities of the echolocating bats could be confirmed.

It was delightful staying at 'The Stables'. We could work during the day at tables in our cabin or on its *khondi*, and we were usually watched by Nido, the female Sable Antelope who had been hand-reared by the Burgesses. Nido was free to roam, but she thought she was a horse and usually stayed in the horse paddocks. One day, I went horse-riding with Jayn and we were followed by Nido who still enjoyed outings with the horses although she was now quite elderly. It was very amusing. I was following Jayn and being followed by Nido, so when Nido thought we should be going faster, she simply poked her long, curved horns into my horse's rear end! You can imagine the horse's reaction! Later, when Nido thought we had gone far enough, she crashed through the undergrowth and came out on the path in front of Jayn. There she stood, facing the horses and blocking the path. There was no alternative – we had to turn around and go back! Just what Nido wanted!

During the year, we had five trips to Liwonde National Park, the first being with Jonathan and Lena in early December. It was great to be back in this place that had meant such a lot to us during 1984–85, but we found many changes. These included a new youth hostel that had been built with money from the World Wildlife Fund. The hostel had a large dining room (also used for classes and lectures) with a kitchen attached, and there were six rondavels each having four bunks, a loo that could be flushed, and a shower. Solar panels provided twenty-four-volt power, so there were electric lights as well. We were allowed to use two of the rondavels and this was a luxury after the accommodation we had had in the past!

We set nets and did some 'batting' in the vicinity of the youth hostel and also around the game scouts' camp where we had done so much 'batting' in 1984–85. One day, the four of us were able to catch several *Taphozous mauritianus* that were roosting on the walls of the new buildings at the hostel. We also set nets over the Likwenu River again, and this time caught twenty-two bats representing five species. And, of course, we looked in some culverts and Jonathan, for old times' sake, crawled through one and caught a *Nycteris macrotis*!

Our second trip was in mid-February and, again, we stayed at the youth hostel. 'Batting' around the hostel was very productive and we added *Pipistrellus zuluensis* and *Pipistrellus stanleyi* to the list of species we needed this year. We had another look in the hollow baobab where we had found *Nycteris grandis* in 1985 and caught one, and that was also very pleasing because we badly needed wing tracings of this species too. We went to the Likwenu River again, and this time netted two large *Macronycteris vittatus* and tape-recorded their echolocation calls while they were hand-held. One evening at Liwonde National Park, we saw a very small bat flying around the dining-room and I went after it with a butterfly net. It was an amazing bat, which fluttered around quite slowly, but it dodged the net time after time! However, I did manage to get it in the end. It was a *Nycteris hispida*, a new record for the park and another valuable source of flight-test and wing morphology data.

Our penultimate trip was in early May and our first night was very exciting because we netted our first ever *Kerivoula argentata* in a net set near the youth hostel. We also caught *Rousettus aegyptiacus*, *Rhinolophus fumigatus*, *Taphozous mauritianus*, *Nycteris macrotis*, *Tadarida condylura*, *Tadarida pumila*, *Pipistrellus zuluensis* and *Nycticeinops schlieffeni* that night and by the time we had made tape-recordings and flight-tests, and had light-tagged two *Nycticeinops schlieffeni* and the *Rhinolophus fumigatus*, it was nearly midnight and we were exhausted! The fruit bat, *Rousettus aegyptiacus*, was the first we had ever caught, and we kept him for a few days. He loved bananas and ate one or two every day, so the bottom of his bag was soon filled with very messy, moist droppings. Consequently, he had to be 'changed' up to four times every day and the washing line was soon adorned with yellow 'nappies'. It was just like having a baby! We also spent a lot of time watching several *Taphozous mauritianus*, including a mother with a juvenile, who were roosting on the walls of the buildings. We had netted the juvenile and, when it was released near where two adults were roosting, it began to vocalize and one of the adults answered. Then the juvenile scrambled up the wall to what was obviously its mother. She nuzzled and licked the baby, and then it pushed under her and sucked for about five minutes.

The next day, we went to the office at the entrance of the park to arrange for a game scout to come with us to the Likwenu River that night. We had a very busy night at the river because we netted twenty bats representing eight species. There were many insects over the river, so the bats were undoubtedly coming to feed as well as to drink. We heard fruit bats calling from the trees near the river – the first time we had ever heard them at Liwonde National Park.

Our first trip to Lengwe National Park since June 1985 began on 17 November, and Jonathan came with us. The escarpment still looked denuded although eucalyptus plantations had been established in some areas. The Lower Shire Valley was looking greener than we expected for mid-November. At Lengwe National Park, we stayed in a new chalet, but it was badly designed and poorly furnished, and we felt it was a great shame that better use was not made of the park as a tourist attraction. We spent several hours at the main hide and saw Buffalo, Nyala, Lesser Kudu, Bushbuck, Warthogs, Bushpigs, Blue Monkeys, Vervet Monkeys and Yellow Baboons. Jonathan saw a herd of eighty-five buffaloes that completely filled the water hole!

We set the harp-trap to catch *Tadarida* emerging from the toilet-block, and it was very successful. And, one night, we set mist-nets around the chalets and although we only caught five bats, one of them was a *Myotis bocagii* – a new record for Lengwe National Park.

Exploring pastures new

Ntchisi Forest Reserve

The day after Jonathan arrived in Malawi in October, all three of us set off for Ntchisi Forest Reserve on the eastern edge of the Central Plateau. This reserve includes Ntchisi Mountain (1642 metres above sea level), which is one of the isolated mountains in Malawi with submontane seasonal rainforest covering its summit. The reserve also includes open-canopy miombo woodland, dominated by *Brachystegia* spp., which surrounds the rainforest. Outside the reserve, there is very poorly managed subsistence farmland that is badly eroded and very depressing. Furthermore, although the forest reserve is supposed to protect the rainforest and the woodland, the trees were being cut down for timber and firewood, and the farmlands were rapidly encroaching onto the reserve. We were also sorry to see extensive pine plantations in this reserve. We felt that Ntchisi Mountain, as an important refuge for wildlife, was seriously threatened, and that there was a very urgent need to survey its bats and small mammals and to do all we could to have it better protected.

Ntchisi Forest Reserve is about sixty kilometres from Lilongwe – and the last sixteen kilometres took two and a half hours because the road was very bad. We were very relieved to reach our destination – the rest-house that had been built by a district commissioner in 1914 as his summer retreat. It was a delightful place in which to stay. It had several spacious bedrooms with comfortable beds and bushlights, and each bedroom had its own shower, basin and loo. There was also a large living room furnished with wooden tables, chairs and tall lampstands, and there was a Tilley pressure lantern that gave a wonderful light, and several kerosene lamps. There was also a night watchman who made sure that a Rhodesian boiler was well stoked, so we had hot water for showers. A real treat! We originally intended to stay for three nights, but we had so much 'batting' to do that we stayed on for an extra night.

During our first night, we set mist-nets along, or across, paths running through the miombo woodland near the rest-house and left them open for two hours. We only caught eight bats, but they represented seven species. These included a horseshoe bat that we first identified as *Rhinolophus simulator* but then, when we analysed its echolocation calls, we found that the CF component was 102–104 kHz whereas that of *Rhinolophus simulator* is 84–86 kHz. Later, when we examined the noseleaf of this bat under a microscope, we were able to identify it as *Rhinolophus swinnyi*. It was a new species record for Malawi and its presence on Ntchisi Mountain was ammunition for a campaign to have the natural habitats in this reserve conserved and protected.

The next day, we drove up a narrow road until the miombo woodland gave way to the rainforest that capped Ntchisi Mountain. It was cool and shady in that forest. There were giant trees with buttress roots, and strangler figs whose roots dropped down to the ground from a great height. We crossed a small stream and found some places under the trees where mist-nets could be set, and we went back in the late afternoon to set them. We placed two across the road, one across the stream, and one in the forest. Bats were plentiful and, after two hours, we had netted twenty-one individuals representing six species. We also had another evening of netting in the miombo woodland near the rest-house and added four more species to the list from this habitat. All in all, our Ntchisi records comprised fourteen species and all were new Malawian locality records.

David also set live-traps for rodents and shrews in the miombo woodland (where he caught three species of rodents) and in the forest where he caught nothing except for one *Beamys hindei* – but the *Beamys* was very special! Some local children also brought us three more species including a very lively mole-rat that snapped its protruding teeth at us and made a clatter of short, loud vocalizations, each of which jerked its whole body. It was a feisty little animal that we named Scissors. As we had done with other mole-rats from Zomba in 1984–85, we put Scissors into a bucket with earth and leaf-litter, and we took him on to our next destination before taking him with us back to Zomba. But, unlike the other mole-rats we had kept in captivity (see Chapter 9), Scissors never settled down and always greeted us with further displays of fury and rage. Nevertheless, I wondered if he would respond gently to my finger, as the other mole-rats had done. He didn't! We subsequently found out that Scissors was a different species of mole-rat – *Cryptomys* sp., not *Heliophobius argenteocinereus*, the species we had at Zomba. Perhaps this explains the difference in his behaviour.

A very special, real mother bat

One night, when we had set nets in the miombo woodland near the rest-house, we caught two fruit bats. One was an *Epomophorus wahlbergi*. The other was probably the same species, but I did not examine her closely enough to confirm this. She was an adult female and I was horrified when I first saw her – she was dangling with her head uppermost and both wings and both feet entangled, and beside her was a tiny baby that was far too young to be separated from its mother. Its eyes were still closed and its downy fur was still too short to keep it warm. I took off my thick leather gloves because I knew I could not disentangle the bats carefully enough with gloves on, but I knew I risked being bitten. I approached the bats extra slowly, talked softly to them, and just *willed* the mother to stay still. And she *did* stay still! She was unlike any other fruit bat I had ever netted. Usually, they struggle, bite and scream loudly, but this one stayed so still that I was able to make her more comfortable by turning her 'upside-down' and then I freed her left wing. Next, I extricated the baby, tucked it under the mother's free wing and was greatly relieved when it attached itself to her nipple and grasped her fur with its well-developed hindfeet. And the mother continued to stay still although I could see that her heart was pounding. I held the left wing firmly over the baby and started trying to free the right wing. It was difficult but eventually I was able to free it and wrap it, too, around the baby. Disentangling the feet and toes is usually very difficult because the terrified bats hang on to the strands of the net with all their strength, but it seemed as though this one sensed that I was getting her out of trouble – she stayed relaxed and made no attempt to bite me. At last, she was disentangled but I did not dare to let her go immediately in case she flew back into the net, so I held her against my chest with one hand while letting her cling to a finger of my other hand. I could feel her heart pounding. Then I began walking very slowly away from the net and, by the time I was well away from it, the mother's heart rate had returned to normal. Then I eased my hold on the mother and moved the hand from which she was clinging until she was almost at arm's length. I expected her to fly immediately but, instead, she looked around, sniffed the air, and seemed to be investigating, with evident curiosity, the very strange situation in which she had found herself. I am sure she was no longer afraid, and I think she perceived me not as a predator but as the one who had helped her escape from the net. The net was the enemy, not me. After a while, I wondered when she would want to fly, and I had other nets to check, so I walked to a small tree and manoeuvred her against a twig so that she had to cling to it. Then I started to back away and only then did she take to the air and fly away with her baby.

Nkhotakota Game Reserve

After Ntchisi Forest Reserve, we went to Nkhotakota Game Reserve. Nkhotakota Game Reserve conserves miombo woodland on the steeply undulating hills and plains around the Bua and Kaombe rivers. As well as the dominant *Brachystegia* trees, there are *Terminalia*, *Kigelia* and other trees and many *Raphia* palms, and this reserve was famous for its large herds of elephants. We stayed at Chipata Camp where the Wildlife Society had built accommodation for students. There were four mud-brick and thatch dormitories with bunks, a kitchen, a toilet-block and a huge dining-room. There were also two delightful thatched rondavels, each with twin beds, which we were allowed to use. These did not have mosquito nets because the camp was too high and too cool for mosquitoes. The camp is located on the side of Chipata Mountain, which reaches 1638 metres above sea level, and it had a wonderful view over *Brachystegia* trees (clothed in pale green, dark green and coppery-red leaves), chiwale palms and dambos far below. Further away, we could see tree-covered hills and then, almost lost in the haze, Lake Malawi.

We had two attendants at the camp – Michael and Raphael. On the first afternoon, David, Jonathan and Raphael clambered up to the summit of Chipata Mountain, which, like Ntchisi Mountain, had submontane seasonal rainforest at the top. However, unlike Ntchisi Mountain, there was no ecotone between the forest and the short grassland lower down – the forest ended very abruptly. The track David and the boys followed was littered with elephant dung and the droppings of Lesser Kudu, Bushpigs and Yellow Baboons. They found it very open under the trees – not much undergrowth and many signs of elephants. Nevertheless, they set live-traps in the forest, and then two more lines of live-traps in the grasslands nearer to the camp. The catch was disappointing. In the forest, there was only one *Mastomys natalensis* and one *Acomys spinosissimus*, neither of which was expected in that habitat. Trapping in the grasslands was also disappointing.

The 'batting' was much more rewarding. We could only set mist-nets close to the camp buildings – there were too many dangerous animals further afield – but we netted *Epomophorus labiatus*, *Rhinolophus blasii*, *Nycteris thebaica*, *Taphozous mauritianus* and *Laephotis botswanae*. We also set the harp-trap over a spring-fed pool in a small patch of evergreen forest nearby, and we just hoped that elephants and monkeys would not come to the pool to drink and then decide to play with this strange contraption. The harp-trap proved a great success. The next morning, safely tucked up under its flap, there was a *Hipposideros ruber*, five *Pipistrellus hesperidus* and four small *Miniopterus*, and on the morning after that, there were four more *Pipistrellus hesperidus* and five more *Miniopterus*. We spent a lot of time flight-testing these bats and recording their echolocation calls. Of the three

Taphozous mauritianus that we caught, one was an adult male in reproductive condition, one was a lactating female and the other was a volant juvenile female. We put all of these bats into separate cloth bags, but soon the adult female and the juvenile started calling to each other with calls that were audible to us as well, so we put them together and the juvenile almost immediately started to suck from what was evidently its mother.

We had some delightful walks from the camp – always accompanied by Raphael. Each morning, we walked up to the forest near the summit and found it swathed in mist. There were Blue Monkeys in the trees and baboons on the ground and, one day, a herd of female Lesser Kudu were grazing just below the forest. One afternoon, we walked downhill from the campsite and saw a huge fissure that had opened up during an earthquake in 1989. From there, we walked along a stream through lots of chiwali palms, dambos and miombo woodland. We saw droppings of many animals but only saw two female Lesser Kudu. It was overcast that day and not unpleasantly hot. Raphael led us up again to a very rocky area where there were huge boulders with caves and fissures. According to Raphael, there were many Spotted Hyaenas living among the boulders, and we found some of their droppings – these are characteristically very hard and almost pure white because of the calcium in all the bones that hyaenas digest. We heard the weird rising 'whoooop' calls of these hyaenas every night – as well as the trumpeting of elephants.

Thyolo (formerly Cholo)

While researching the bats and small mammals of Malawi in 1984–85, David and I often came across an unusual citation in the literature – 'Wood in Kershaw 1922'. This was a paper by Mr P. S. Kershaw who worked in the mammal section of the Natural History Museum in London. The title of the paper was 'On a collection of mammals from Chiromo and Cholo, Ruo, Nyasaland made by Mr Rodney C. Wood, with field notes by the collector'. Rodney Wood had collected twenty-eight species of bats and twenty-seven species of rodents, shrews and sengis (elephant-shrews) and five other larger mammals, most of them from Chiromo. They included some of the bats that appear to be the rarest in Africa although perhaps this only means that they are the most difficult to catch. They also included a rare mouse from Thyolo that was originally named *Uranomys woodi* after

Rodney Wood, but is now known as *Uranomys ruddi*. When we returned to Malawi in 1993–94, we were especially keen to visit these two localities in case we could also collect some of these 'Wood specials'.

We had friends who were renting a large colonial-era house ('Magombwa House') on the 'Kasemberika' tea estate near Thyolo (as Cholo is now spelled), and they invited us to stay there in January and again in March. Thyolo is a very lovely area near the base of Mount Mulanje. We drove through it on our way to Mount Mulanje in 1985 and it was very enjoyable seeing again the rolling hills with verdant, emerald green tea plantations sheltered by Australian *Grevillea robusta* trees, and the patches of relict lowland rainforest and riparian forest in valleys that were too steep for growing tea. When Rodney Wood made his collections from Thyolo between 1919 and 1921, most of the area now under tea or tobacco was miombo woodland, but there are still some patches of this habitat left. 'Magombwa House' was built of mud-bricks that were hand-made and fired, and it had a corrugated-iron roof, polished concrete floors, and the doors and window frames were made of beautiful local hardwoods. Its many rooms were large and had high ceilings, and their windows opened onto wide *khondis* with mud-brick pillars supporting their roofs. The house was surrounded by several hectares of ornamental garden with many old trees, palms and colourful flowers of many kinds. 'Magombwa House' had a wonderful atmosphere and we especially enjoyed sitting around a large log fire at night, swapping yarns with our friends.

We mentioned that we were particularly interested in Thyolo because of Rodney Wood and you can imagine our reaction when we were told that it was he who had built 'Magombwa House' and lived in it! I was excited by this coincidence, and suggested to David that, if we obtained any of the 'Wood specials', we should write a paper called 'Thyolo and Chiromo Revisited' in which we could compare the environment in Wood's time (prior to 1922) with that of today (seventy-two years later).

'Magombwa' proved disappointing with respect to bats, and we did not catch any of the 'Wood specials' so we moved to another Thyolo tea estate – 'Satemwa' – owned by Chip and Dawn Kay. 'Satemwa' is a huge, very successful estate on undulating country with vast plantations of tea, patches of forest and dams in the steep-sided valleys, and some eucalyptus and pine plantations. In the distance, we could see the shallow dome of Thyolo Mountain covered with dark green submontane forest. We had been warmly invited to stay for four nights, and we were woken up every morning by a servant who brought us cups of 'Satemwa' tea!

On our first evening, the Kays' son, Hilliard, took me to a series of fishponds where bats were flying. I made tape-recordings of their echolocation calls and analysed them the following day. The calls included those of *Myotis bocagii*, a species of *Scotophilus* and *Pipistrellus nanus*, so David and I went back to one of the ponds at dusk the next day and set mist-nets. The results were fantastic! We netted fourteen *Myotis bocagii*. These bats are exquisitely beautiful vesper bats with rufous dorsal pelage, cream ventral pelage and blackish noses, ears and flight membranes. We watched them swooping very low over the dam and made more recordings. We also netted one *Epomophorus wahlbergi*, one *Rhinolophus blasii*, three *Scotoecus hirundo* (a new record for Thyolo), one *Scotophilus dinganii*, six *Pipistrellus nanus*, two *Pipistrellus rueppellii* and, best of all, one African Trident Bat (*Triaenops afra*) – a species not previously recorded from Malawi and not often recorded far from the East African coast! I could not believe my eyes when I saw this bat in the net! It appeared to be made of burnished coppery gold – this being the orange-phase of a species that also has a grey-phase. I had never seen a living *Triaenops* before, but I recognized it instantly because it has a very characteristic noseleaf with three long, trident-like projections rising from its posterior edge.

One afternoon, we visited 'Namingomba' estate, which lies next to 'Satemwa', and the general manager drove us through tea plantations to a patch of chiwali palms (*Raphia farinifera*). Chiwali palms have the longest leaves of any terrestrial plant, but what interested us most was that a large colony of African Straw-coloured Fruit Bats (*Eidolon helvum*) was roosting in these palms, and this was the first time that we had seen these large bats in Malawi. We watched them through binoculars. They were very active, clambering around the palm fronds using their thumbs and forearms as well as their hindlimbs. There was much jostling, climbing over each other, grooming of themselves and each other, and chattering, and we wondered if they ever went to sleep.

On our last day at 'Satemwa', we investigated a colony of bats roosting in the overflow tunnels leading away from another large dam. We set mist-nets across the entrances of the tunnels, and then someone walked into them to flush the bats out. In this way, we obtained twenty-one *Miniopterus* and one *Rhinolophus clivosus*. We got good tape-recordings of the *Miniopterus* as we let most of them go to fly back to the tunnels. All this took some time, and we did not leave 'Satemwa' until three o'clock and therefore did not get to our next destination until dusk.

Our next destination was 'Mbala', the home and tobacco estate of our dear friends Don and An Pyman. It was Easter, so we were unable to stay at 'Kapalasa' because Sven's family was visiting and needed 'our' room. We were very grateful that the Pymans could put a roof over our heads for four days, and we took the opportunity to set mist-nets over their swimming pool. We also went back to 'Kapalasa' to conduct a Banana Pipistrelle census on one day. Then it was time to head off for the second of the Rodney Wood collecting localities.

Chiromo and the most exciting night of 'batting' we ever had

Chiromo, in the Lower Shire Valley, was the second of the two localities we were especially keen to visit because of Rodney Wood who had lived there between 1914 and 1920. He had a house on the bank of the Ruo River, which joins the Shire River at Chiromo. His house was on a cotton plantation surrounded by dense riverine forest. That area is flat, but the escarpment leading up to the Shire Highlands is not far away, and Elephant Marsh – a huge swamp – is also nearby. Wood recorded that *Hyphaene* palms were particularly abundant where he lived, and that they were associated with several species of fig trees (*Ficus*) and other trees, many of which were very large. Also, in some areas of more open woodland bordering the rivers, and in Elephant Marsh, there were *Borassus* palms. All of these trees provide domiciles for bats. Rodney Wood reported *Taphozous mauritianus* roosting on the trunks of large trees, *Nycteris hispida*, *Scotophilus dinganii*, *Myotis bocagii* and *Tadarida pumila* in hollow tree trunks, and the apparently very rare *Scotophilus nigrita* in hollow trunks and large holes in the *Hyphaene* palms. He also found the rare *Scotoecus albofuscus woodi* (named after him) among the leaves of young, low *Hyphaene* palms. It was not surprising that the bat fauna of Chiromo was rich. The vegetation was varied and provided fruits and roosts, the climate was warm throughout the year and insects were always abundant. Altogether, Rodney Wood collected twenty-six species of bats from the Chiromo District and, as well as those mentioned above, these included *Nycteris woodi* (also named after him), *Eptesicus hottentotus*, *Glauconycteris variegata*, *Pipistrellus rendalli* and *Myotis welwitschii* all of which are listed as 'rarely-recorded' in Malawi because, prior to 1997, no more than ten individuals had been recorded. These species are also considered to be rare throughout their ranges in Africa. Furthermore, within Malawi, three of these species have only been found at Chiromo and two are known from only one or two other localities in Malawi. Rodney Wood's collection from Chiromo also included three species that are 'seldom-recorded' in Malawi (i.e. only eleven to twenty individuals had been recorded prior to 1997). We had never caught these species, so no wonder we were keen to go to Chiromo!

We left on 7 April with the Transporter heavily laden and drove through Blantyre and down the escarpment as though we were going to Lengwe National Park. After the Lengwe turnoff, the road deteriorated badly. It was a mosaic of old tarmac, stones and bare sand and we seemed to be climbing in and out of awful potholes all the way. We were in second or third gear, and it was very hot. The crops – mainly millet here – looked very poor, and the villages looked neglected and semi-abandoned.

Chiromo, which had once been a thriving, bustling port on the Shire River, had been periodically devastated by floods and in 1989, after a particularly severe flood, it was abandoned and most of its residents moved to higher ground and established the settlement of Bangula about five kilometres away. We had arranged to stay in Bangula, in a house owned by Cotton Ginners Ltd., and we were delighted to find that it had electricity, water, overhead fans, beds with clean sheets and a refrigerator with cold drinks. All the *khondis* and windows had flyscreens, and the house was sheltered by shady neem trees.

The next morning, we set off to explore what remained of Chiromo. We drove along a dusty laterite road that followed the old railway line. Sand had filled the spaces between the sleepers, and we saw people walking or riding bicycles along the railway because it was much less bumpy than the road! On the right, leading down to the Shire River, there were extensive flats with patches of low shrubs, palms and reeds, and large pools with mats of flowering water hyacinths around their edges. We crossed the Shire by means of the railway bridge – we had to straddle the railway lines! It was an amazing bridge with steel girders that had been built in the late 1940s, and it had a path for pedestrians and animals on one side. A little further on, we drove along an avenue of old trees and soon passed all that remains of old Chiromo – brick buildings, mostly lacking roofs, doors and windows but still shaded by old trees. We drove through Chiromo and checked culverts under the road and captured one *Nycteris macrotis*, and we captured one *Nycteris grandis* (a new record for Chiromo) in the ruins of the old district police station that were now carpeted with goat droppings. Although Chiromo is now a ghost town, it was potentially a wonderful place for bats.

As we drove back towards the bridge, we stopped near an old gateway with stately pillars on either side, and we walked through into the remains of a large ornamental garden that we hoped extended as far as the steep banks of the Ruo River. The confluence of the Ruo River with the Shire River was just south of the bridge. The garden *did* run down to the river, and it surrounded the remains of a stately, two-storied house that had been the home of the manager of the British Cotton Growing Association in the old days. The house and grounds were guarded by a night watchman who lived in a small cottage at the edge of the garden. We approached the cottage to ask if we could come back at night to catch bats, and we were greeted by a crowd of excited children who had probably rarely seen white people before. We were also greeted by one of the most attractive Malawian dogs that I had ever seen! She was lithe and graceful with short cream fur, pricked ears and a long feathery tail. There was no artificial selection in this dog's breeding – only natural selection – so she looked exactly as dogs are meant to look like in this part of the world! I wished I could have brought her back to Australia.

Having obtained permission to return to this garden, we came back in the late afternoon. As Rodney Wood had done, we found *Taphozous mauritianus* roosting on the trunks of some of the trees, and we obtained some very useful tape-recordings of their echolocation calls when they flew from their roosts and began foraging in the open sky overhead. Then we set mist-nets in the garden and across a gap in the trees lining the bank of the Ruo. There was no moon and only the lightest of breezes coming off the river, so conditions were perfect! While we waited for dusk to fall, the night watchman cooked his evening meal over a tiny fire out in the open, and a fisherman walked home carrying two African Lungfish, which we were very interested to look at. And then the first bats began to fly and we waited, very tensely, near the nets.

It was the most exciting night of 'batting' that we ever had! At first, we netted many species that we were familiar with, but then, one by one, we started catching 'Wood specials' that we had never seen before – *Pipistrellus rendalli* and *Scotoecus albofuscus woodi*, both of which have white wings, and *Glauconycteris variegata*, the beautiful Variegated Butterfly Bat, which has pale fawn pelage, and yellowish wings with dark brown venation that look just like dead leaves. In total, we netted twenty-five bats representing eight species, and it was one o'clock in the morning before we had finished checking them, testing their flying abilities in the $1 \times 1 \times 1$ metre enclosure, making descriptions of the species we had not seen before, and making wing tracings.

The second day was also busy and rewarding. We enlisted the help of some young men who lived in a village a few kilometres south of Bangula. They climbed under the corrugated-iron roof of a house and caught three *Tadarida condylura* and one *Tadarida pumila*. They also showed us a colony of fruit bats that appeared to be *Epomophorus labiatus*, but we were unable to catch any to confirm the identification. That night, we went back to the old house near the confluence of the two rivers, but only netted four bats. The word had got around it seemed! However, one of the bats was *Epomophorus labiatus* and this was another new record for the district.

On another evening, we set mist-nets over the swampy area between the road to Chiromo and the Shire River. We could see many bats flying over the pools and over the low vegetation and feeding on small flying insects that were very plentiful. We recorded the echolocation calls of these foraging bats but we only netted one. It was *Pipistrellus rendalli* and its echolocation calls indicated that most of the bats we were watching represented this species too.

In total, we captured fifty-four bats representing fifteen species, half of the thirty species recorded from Chiromo either by Rodney Wood, other bat biologists, or by us on this trip. Three of the species that we caught were new records for the district. Of course, ours was only a four-day survey whereas Rodney Wood spent six years in this area, and therefore we could not conclude that the species we did not catch were no longer there. What was interesting was that at least eleven of the species that Rodney Wood recorded *were* still there despite all the environmental changes that had taken place since his time.

After this trip to Chiromo, it was clearly time to write *Thyolo and Chiromo Revisited* but we realized that we actually knew very little about Rodney Wood – and neither, it appeared, did anyone else in Malawi! David took on the challenge of finding out everything he could about this mystery man and, after spending many intriguing years talking to people who had known him, visiting places in Britain, Zimbabwe and the Seychelles where he had lived, researching old archives in Malawi and elsewhere, finding his diaries in the Commonwealth Library in Cambridge, and reading many books about Nyasaland in the old days, he published *African Naturalist – the Life and Times of Rodney Carrington Wood, 1889–1962*. We certainly had no regrets about visiting Thyolo and Chiromo!

In the end, although the first six months were very difficult, the last six months were wonderful and very productive. We returned to Australia, very satisfied, to write up the results of our research and to return to teaching zoology.

12

ONE-SPOT, TWO-SPOTS, A BORED FISH, A MISCHIEVIOUS DOLPHIN AND A HERRING GULL

One-spot and Two-spots: two very different Hooded Rats

When we returned to Canberra after our second trip to Malawi, as well as collaborating with David to write papers, I ran a course on animal behaviour for the Australian National University's Centre for Continuing Education, and this was great fun and very interesting. I designed a series of practical experiments to help the students understand some of the intricacies of animal behaviour and get some hands-on experience, and I also gave some demonstrations of animal behaviour and these included some aspects of the behaviour of two female Hooded Rats that I bought from a pet shop for the purpose and kept at home. I learned a lot from those rats – in particular that they differed enormously in temperament. One demonstration involved watching and recording exploratory behaviour as each rat was placed in a 2.5 × 1 metre arena. One-spot, the bold one, explored the arena much more enthusiastically and comprehensively than Two-spots who appeared to be very timid.

I also designed and built a 'puzzle-box' that was about one metre high, one metre long and fifteen centimetres wide, with a series of five 'shelves' with at least one puzzle to be solved on each level. A rat was released through a door opening onto the top level. To escape from the box, the rat had to (1) open a sliding door on the top level, (2) go to the end of the level and then climb down a wire-mesh wall to level five at the bottom, (3) go under the bottom of the mesh and then climb up the other side to level two (although all levels were accessible from the mesh, they led to closed doors or dead ends at this stage), (4) go over a see-saw on level two, which raised a bolt that had been preventing a hanging door on level four from opening, (5) proceed to end of level two and go through an open hole in the floor to level three, (6) push open a door hanging down from the roof, then proceed to end of level three and climb down the mesh to level four, (7) push through a hanging door that could now be opened (as a result of going over the see-saw on level two), (8) go to the end of level four and slide open a trapdoor in the floor and (9) go through the trapdoor to the left half of level 5 and then leave the box through an open door. (Note, the right half of level 5 could be accessed when the rat had first climbed down the mesh, but it was sealed off and therefore did not lead anywhere.) Having escaped from the box, the rat was immediately returned to her home-cage.

The time taken to reach the exit was recorded. Because rats feel distressed in foreign places, both rats were initially highly motivated to escape and return to their familiar home-cage. But their different temperaments were very apparent. The bold rat took approximately eleven minutes to get out of the puzzle-box on her first attempt, then she took sixteen minutes, eleven minutes, four minutes, three minutes and then two minutes on trials two to six. By that time, she knew exactly what route to follow and how to negotiate every puzzle without any loss of time. However, from trial six onwards, she took longer and longer to get out of the puzzle-box and the explanation was that the box had become familiar territory in which the rat evidently enjoyed playing with the puzzles. In contrast, the timid rat never found the courage to climb down the wire-mesh, so she remained on the top level until I rescued her.

I also made a 'Skinner box' for these rats. This was a simple box with a lever that a rat could press down to release a pellet of food and, at first, a pellet was released *every* time the lever was pressed. Both rats learned, by trial and error, how to obtain their food in this way, and they would press the lever when they were hungry and stop pressing it when they were satiated. Then I adjusted things so that pressing the lever sometimes delivered a pellet – and sometimes did not. Delivery or otherwise was random. And the behaviour of the rats was transformed! They literally became gambling addicts who spent all their time in the Skinner box pressing the lever, looking to see if a pellet was delivered and then pressing the lever again and again and again *ad infinitum*, whether or not they were hungry. It was a very sobering lesson.

A fishy solution to boredom

When we returned to Canberra, I also continued to be a demonstrator in 'Humans and Vertebrates' and, as one of the practical sessions, the course lecturer and I took students to the National Aquarium in Canberra to observe the many various functions of the fins of fish, and to do a project on fish from the Great Barrier Reef. However, what I remember best was the behaviour of a large, very bored wrasse from New Zealand. This fish lived with many other fish in a very large, cylindrical glass tank that was the height of two floors of the Aquarium. It was located near the main entrance, and many schoolchildren loved to touch the glass tank with hands that were seldom clean. Consequently, one of the staff had to clean the glass, very frequently, with some sort of spray that he wiped off with a large cloth. I was interested that the wrasse swam up to the glass and chased the cloth as it swished to and fro in front of its eyes. I commented about this to the cleaner who said, "Oh yes. It always does this. But watch what happens when I stop." What did happen? The wrasse swam to the top of the tank and, using its head, splashed a lot of water over the edge of the tank so that it trickled down the outside and made the glass dirty again. The obliging cleaner went to work again, and again the wrasse chased the cloth.

Animals of all kinds can become very bored in captivity – even fish.

Our data from Malawi

Although the first half of our second expedition to Malawi was fraught with troubles and problems, the second half was unbelievably productive and we returned to Canberra with a huge amount of data to analyse and many papers to write. We soon published six papers and then I began analysing and integrating of all the data we had collected on the echolocation, wing morphology, flight, diet, foraging behaviour and domiciles of the forty-two species of microbats that we had found in the bat communities on Zomba Plateau (approximately 1800 metres above sea level) and at 'Kapalasa' Farm

(1000 metres), Zomba (800–900 metres), Liwonde National Park (500 metres), and Chiromo District (less than 100 metres). We decided that our information about this resource partitioning in the five communities of microbats that we studied was worth putting into a talk to be given at the 7th Symposium on African Small Mammals, which was held in the Itala Game Reserve in South Africa in late September 1995.

"I can play tricks too" – a very mischievous dolphin

We went to South Africa about a month before the symposium at Itala. First of all, we spent some time working at the Transvaal Museum in Pretoria where there are good collections of bats and rodents. Then we flew to Port Elizabeth on the south coast, hired a car and began a journey along the beautiful and well-known Garden Route to Stellenbosch. This journey enabled us to spend time with colleagues who were working on rodents or bats or other mammals of interest, and these kind people took us to visit some national parks, game parks and some other places where they were carrying out interesting research.

At Port Elizabeth, we spent some time with a zoologist who was studying seals. She had arranged for us to have a tour of the 'Bay World' oceanarium and this gave me my first close-up encounter with two Bottle-nosed Dolphins who were kept in a very large outdoor enclosure.

The dolphins saw us approaching their pool with two of their trainers, and they swam over to greet us. One of the trainers had a ball that she threw to one of the dolphins, and the dolphin caught it in its mouth. Then it threw the ball, very accurately, back to the trainer who caught it and threw it back to the dolphin. This game of 'catch and throw' continued for a few minutes, and then I asked if the dolphin would play the game with me. The trainer gave me the ball and I threw it to the dolphin who caught it with its usual expertise. Then it threw it back – but just far enough to my right that I could not catch it! Having just seen how accurately the dolphin could throw the ball back to its trainer, I had no doubt that the missthrow was deliberate and mischievous, and I am equally sure that the dolphin enjoyed my discomfort! I retrieved the ball and tried again and, this time, the ball was thrown back to my left, and again just too far for me to catch it!

Everyone was highly amused by this mischief, but the trainer had some tricks up her sleeve too. She took us down some stairs so we could look through a huge plate-glass window to see the dolphins under water, and we could also listen to their echolocation calls through a special microphone. The dolphins could see us with their eyes, but they could not perceive us with their echolocation calls because these were reflected back by the glass window. Then the trainer threw the ball towards a dolphin who could see it with its eyes, but we also heard the burst of sound as the poor beast also tried to 'watch' the ball with its echolocation calls. Of course, the ball bounced back off the glass, so the dolphin could not catch it – and this must have been very frustrating indeed!

After a few more throws that bounced back off the glass, we all went upstairs again, and the trainer tried to pacify the frustrated dolphin by throwing the ball to it again. Did the dolphin throw it back? No! It threw it right across the pool and it landed inside some pens that the trainer could only get to, with considerable difficulty, by balancing on the very narrow walls of the pens! Undoubtedly, that dolphin had the last laugh!

The road to 'Mammals of Africa'

We presented our paper on resource partitioning in bat communities in Malawi at the 7th International Symposium on African Small Mammals at Itala Game Reserve in South Africa in 1995. And then I continued analysing data on Malawian bats when we returned to Canberra. However, progress on this project was interrupted by two events. By 1997, the golden years of universities in Australia seemed to be over and the Australian National University was plagued by cuts to funding, strikes and bans, and shortages of technical staff. So David decided it was time for us to retire – and we did so, officially, in mid-August. However, we were able to keep our office in the Botany and Zoology Department and use any facilities we needed. For a while, I continued analysing data and writing papers and David continued researching the life and times of Rodney Wood, but we also wanted to make the most of our new-found freedom by travelling to England, Europe and Africa during the Australian winters.

Our first post-retirement trip overseas was in 1998, and it lasted nearly six months. We visited England and several countries in Europe and then went to the Seychelles to see where Rodney Wood had lived on Cerf Island, and to meet people who had known him. And, finally, we went to Zimbabwe to visit more places where Rodney Wood had either lived or worked, and to meet more friends or relatives who had known him. Both overseas and back in Canberra, we continued to have wonderful encounters with special animals.

A Herring Gull in Scotland

During our overseas trip, David and I hired a car and drove around the east coast of Scotland and then across to Braemar and finally to Nairn where we stayed with some cousins of David's. En route along the coast, we stopped briefly at Stonehaven, a town with a harbour south of Aberdeen. While David had a cup of coffee, I walked across a car park to have a quick look at the beach. Then, as I returned across the almost empty car park, I saw a Herring Gull about fifty metres away, and it was about to be attacked by a Jack Russell Terrier. Instead of flying away from the dog, the gull ran towards me and I realized that it had a broken wing. To my surprise, the gull came right up to me and then folded its wings and settled on top of my feet! The dog came running towards us, growling and barking. I shouted at it and it came to a halt but continued to bark loudly. The Herring Gull, however, seemed to know that I would protect it and it stayed where it was until the embarrassed owner of the dog managed to catch it and drag it away on a leash. I was not sure what to do then. We had no time to try to find a vet and, furthermore, I was not sure I could pick up the bird to take it anywhere – Herring Gulls are quite large and have dangerously strong, sharp beaks. I decided that it had probably been surviving on scraps that people gave it, and was likely to continue doing so, so I moved it off my feet and went back to our car. But I still wonder how that bird knew to come to an unknown human for help.

Gang-gangs surely have a sense of humour too!

My favourite birds are Gang-gangs because they are such characters and their enjoyment of life seems more conspicuous than that of most other birds. They seem to have such fun!

Gang-gangs are the smallest of Australia's cockatoos – their total length is about thirty-five centimetres. Male Gang-gangs have a bright red head with a rather untidy crest of floppy red feathers, but the feathers on their necks and backs are grey with pale grey margins while their ventral feathers are grey with pale reddish or pale grey margins. Female Gang-gangs have a grey head and floppy grey crest, dorsal feathers grey with pale margins and ventral feathers with a mixture of reddish, greenish and whitish margins that make the ventral plumage of females more colourful than that of males.

We have had Gang-gangs coming to our bird tables for many years, and they know that if they make a particular "Ack-Ack" squeaky-grating call, we will put sunflower seed out for them, and stand guard nearby to prevent them being ousted by Sulphur-crested Cockatoos who are much bigger and more dominant. The Gang-gangs have learned that we can wave our hands, flap tea-towels and shout "Shoo!" without there being any need for *them* to panic and fly away from their meal, and they take no notice at all!

We have not known any of 'our' Gang-gangs individually for more than a few months at a time and, if they came back after several months, we were never sure that it was the same ones who came back. However, for many years now, there have been some that would eat out of our hands. And one summer, there was a pair whose four youngsters would fly onto our shoulders or outstretched arms to take seed from our hands. Then their parents became bold enough to land on us and there was one occasion when I had six sitting on me – two on each arm and one on each shoulder. Gang-gangs have very large, strong, wickedly hooked beaks and, at first, I did not completely trust them not to bite a finger instead of a seed but, after a while, I gained sufficient faith to keep my left hand still and let them take a finger in their beaks if they wanted to, and they did – but they were so gentle that I trusted them completely thereafter.

Our first bird table was suspended by four cords from the branch of a tree near the entrance to our *khondi* (verandah). The Gang-gangs could easily fly to the table and land on its rim, but they much preferred the challenge of landing in the tree and then trying to climb – usually head first – down one of the cords. The table and the cord swung wildly, and this game was just sheer fun! Our clothes-line provided a similar challenge. It was almost impossible to stay upright on the sagging line – especially when several birds tried to do this at the same time – and then it was equally tricky to get upright again. But that didn't matter – it was apparently heaps of fun dangling upside-down.

Our neighbours in Spencer Street were very envious of us having friendly Gang-gangs in our garden, and they were all hoping that the birds would come to the bird table when they came to our garden for the Spencer Street Christmas party one year. I told them it was most unlikely that the

Gang-gangs would come when there were so many people and children milling around.

I was wrong! A pair started calling from the huge eucalypt at the edge of the lawn where everyone had gathered, and everyone looked up. This seemed to be the cue for the birds. One of them swooped down and across the lawn, so close to the people that its wings brushed their heads. This was very startling and those in the bird's line of flight hastily stepped back with cries of alarm, and they bumped into those behind them. After that, there were two lines of people with a narrow corridor between them – and the Gang-gangs utilized it by swooping backwards and forwards at about the shoulder-height of the people, brushing faces with their wings as they did so. The corridor widened!

The mischievous birds spent quite a few minutes at this game, and then they went to the bird table and amused everyone by their comical descents of the cords suspending the table. They began to feed, and I encouraged all the children to sit a short distance away so they could watch the birds closely. Then the adults began to crowd around, the children stood up, and the gap between the birds and the people narrowed – but the birds just went on feeding! They made that street party rather special.

13

TRAVELLING WITH SPECIAL COMPANIONS
ALONG A VERY ROCKY ROAD

Great expectations

Soon after we retired, David was invited to be one of three editors of a
definitive, multi-author work – *Mammals of Africa* – to be published by
Academic Press. There were to be six volumes altogether, and David was
to be the editor of two volumes covering the small terrestrial mammals and
bats, of which there are 821 species – 395 rodents, 13 hares and rabbits,
192 small insectivorous mammals (sengies, golden moles, hedgehogs and
shrews) and 221 bats. This meant that he took on approximately 74 per
cent of a total of 1116 species. I volunteered to write a moderate number of
profiles of bats, including Malawian species that I was particularly familiar
with. So, although we stopped publishing papers about our work in Malawi,
most of the unpublished data became incorporated into *Mammals of Africa*.

There were to be profiles on every extant species, and on all of the genera,
families and orders, and they were to be written by the leading experts
on the species involved. Every species profile would include sections on
Taxonomy, Description, Geographic Variation, Similar Species, Distribution
(including a pan-African map), Habitat, Abundance, Adaptations, Foraging
& Food, Echolocation (bats only), Social & Reproductive Behaviour,
Reproduction & Population Structure, Predators, Parasites & Diseases,
Conservation, Measurements and Key References. There were also to be

paintings of representatives of each genus, and pen-and-ink drawings of skulls, teeth and other diagnostic characters. The whole project was an enormous undertaking and, in hindsight, the original plan to publish all six volumes simultaneously within about four years seems ludicrous.

It turned out to be too huge a task for the original three editors to cope with single-handed, so three other editors were added and, from about 2001 onwards, I started editing of all the profiles in the order Chiroptera (bats), which eventually filled about three quarters of Volume IV. There were twenty-two other authors in the team writing the profiles on bats, and I am so grateful for their collaboration, co-operation, tolerance and assistance with the arduous editing process, and for their encouragement throughout long, frustrating delays. The end result was very much a team effort. David and I finished our writing and editing in 2005 but after that, for various reasons, I found myself doing all the pen-and-ink illustrations for the bats and (with David's help) for all the rodents and small insectivorous mammals as well.

The road through *MoA* was a very rocky one. The profiles, including our own of course, needed much editing to make them all comparable and compatible, some people did not deliver the profiles they said they would write (so David and I had to write them) and there were many other problems. So, the project dragged on and on, and went through three publishers altogether before it was eventually published in 2013 by Bloomsbury. However, while doing 'our bit', we had very enjoyable travels to work in museums in Africa, Canada, U. K. and Europe, and we spent time with many of 'our' authors and made many new friends. We also had a dog, the birds and possums in our garden, and some other animals that we encountered during our annual travels abroad who helped to keep us sane. Three immediately come to mind – Moya, Woo and Blossom Possum.

Moya, another much-loved dog

One of these animals included Moya – probably a Labrador–Border Collie cross – who was rescued by our vet son, Jonathan, after she had been dumped in the country, taken in by a kind lady who was trying to find a home for her, and then shot through one front leg by a crank who hated dogs. Finally, she was taken to the veterinary hospital, where Jonathan worked. The crank had also shot the kind lady's dog and getting its life saved was all she could afford. She took her dog to Canberra's top surgeon and then dropped 'Blackie' off at the Canberra Veterinary Hospital to be put down. However, the vets at this hospital decided that two-year-old 'Blackie' had such a lovely temperament that it would be wrong to put her down without attempting to find a home for her. Her wounded leg was treated and then Jonathan brought her to us one morning and told us her story. Poor dog! She was hobbling around on three legs while the other, swathed in bandages, was held high off the ground, and she was wearing one of those Elizabethan ruffs around her neck to prevent her pulling off the bandages. We were about to go overseas for three months, and we had decided not to have another dog after Moshi while we hoped to travel every year. However, this two-year-old dog, who had recently had puppies, won our hearts and, when Jonathan said he would attend to all her veterinary needs and look after her while we were away, we decided to adopt her. We named her Moyali – Moya for short. Moya had to tolerate some extremely painful treatment of her wounded leg, but she never growled, snapped or complained in any way. Often she did not even flinch and, on one occasion while I held her leg while Jonathan pulled the skin over the wound ever tighter and tighter to cover it, I said to him that I wondered if the nerves were so damaged that she could not feel what he was doing. "Oh no," said Jonathan. "She is feeling everything, and it must be agony." It seemed that this very brave dog really understood that she was being helped and loved.

After several setbacks, Moya recovered and eventually she was able to move freely without limping, and Jonathan's very clever treatment of the wound enabled it to heal without leaving a scar or any other trace.

We thought that Moya was a Labrador–Border Collie cross. She looked like a lightweight black Labrador except for a patch of white on her chest, but she did not behave like a Labrador. She was very reluctant to retrieve things or carry them in her mouth, and she did not like swimming. In fact, her behaviour was typically that of a sheepdog and, the first time she saw a mob of kangaroos on 'Dimbilil', she did a perfect cast around them and herded them back to us. But she was also an expert hunter. Within a few minutes of being turned loose at 'Dimbilil', she had caught and killed a perfectly healthy rabbit, and she brought it straight to us. We know she had been lactating not long before we adopted her, and we wondered if she had raised her puppies by hunting.

At home in Spencer Street, we built a kennel for her in the shelter of the *khondi* and she had the run of the back garden at night. One morning, she greeted David at our back door with unmistakable signals that he should follow her. He did so, and she led him to the vegetable garden where we were growing French beans amongst other things. Moya stared at a patch of ground between two bean plants, looked up at David, and then stared again at this bit of ground – dogs point to things by looking at them, looking at you and then looking at them again. To David, the ground looked undisturbed and he was puzzled by Moya's behaviour. And then Moya, ever so carefully, scraped a tiny layer of soil off the ground and then looked at David again. He looked where Moya had removed the sliver of soil – and there was the fur of a possum! Moya was so proud of herself! We clearly understood what she was indicating – "Look what I caught last night! Aren't you thrilled? It's for all of us to share, but it needs to ripen here for a while. In a week it will be *delicious*!" She had managed to bury that poor possum without damaging a single bean plant, and she had put all the soil back so the burial site was invisible. Amazing, and very touching – but we had to chain her up at night from then on because, unlike Moya, we like our possums alive!

We had Moya for thirteen years until she was at least fifteen years old. She was always incredibly sensitive to our moods be they joy, sorrow, anger, anxiety or whatever, and she always gave us the company and comfort we needed throughout those very difficult years with *MoA* problems. However, after a long life, she became very arthritic and it became obvious that she was no longer really enjoying life. We do not believe that dog owners should keep very old and debilitated dogs alive just because they cannot bear to part with them, so – with our vet's total approval – we had her put to sleep. She had a wonderful morning being thoroughly spoiled by us and then the vet came while she was having a nap on her own bed. The vet had some yummy treats with her – so Moya slipped peacefully away with her mouth still full of treats. David and I wish that we could go like that when our time comes.

Woo, a Crested Pigeon

When times are bad, pigeons abandon chicks they are struggling to feed, and they save themselves so they can breed again when food becomes abundant. So it was in 2004 during a long drought.

I was walking to the university and, out of the corner of my eye, I saw a baby Crested Pigeon – one of the native pigeons that occur throughout most of Australia. They are beautiful birds – grey with a pinkish tinge and they have a distinctive, slender black crest that is carried almost vertically above the head. The wings have wavy black bars anteriorly, and an iridescent patch centrally that can look green or purple depending on the angle of view. The tail is long and dark grey with a paler tip and, when one of these pigeons lands on something, the tail flicks forward over the bird's back, giving the impression that the bird is going to lose its balance and topple over. The eyes are bright red. Crested Pigeons spend a lot of time on the ground and they can run very rapidly on their shortish pink legs and long pink toes. Then, when they fly, their flapping wings make a loud whistling noise that may alert other pigeons to possible danger – or perhaps it attracts the attention of predators away from any pigeons that stay on the ground. The whistling sound is made by air rushing over specially modified primary feathers but, during flight, these pigeons sometimes glide silently for short distances. Crested Pigeons have a distinctive call that has been described as 'whoop'. They live in pairs during the breeding season but form small flocks in the autumn.

I almost missed the barely fledged chick on my way to the university. It was crouched against a low brick fence and did not move. There was no sign of its parents, and I wondered if it had been hit by a car on the nearby road. It let me pick it up and examine it. It did not appear to be injured, so I carried it to a safer place on the other side of the brick fence and put it down. I waited for about fifteen minutes, expecting the parents to appear, but this did not happen. However, in case they would come when I was not there, I went into the university and spent two hours in our office before walking home. When I found the little pigeon in exactly the same place as I had left it, I knew I had to intervene if its life were to be saved.

I carried it home and forced some honey-water down its throat before putting it into a cardboard box in a darkened room. Then I went to the nearest pet shop and bought some artificial 'crop milk' – a protein-rich powder that can be mixed with water to make a fluid resembling the crop milk that parent pigeons secrete from the lining of their crops and regurgitate to feed their young. I fed this crop milk to the baby with an eyedropper, and it rapidly regained its strength and began to thrive.

I was hard at work on *MoA* when 'Small' arrived, so I kept him in a box next to my desk until he was able to perch, and then built a perch for him so he could watch me working at my computer. When he got hungry, he would start squealing loudly and I would fetch crop milk for him. Very soon, he began to display as well as squeal when he was hungry – he would throw one wing forward over his head and then waggle it vigorously before withdrawing it. Then he would waggle the other wing, and then alternate from one wing to the other. It was not long before Small could fly from his perch onto my shoulder and then, when he was hungry, he would thrust his bill into my ear, squeal excruciatingly loudly, and waggle his wings at the same time. This undoubtedly resulted in many typos in the *MoA* manuscript!

Small was great company during those weeks of writing and editing, but he could not stay in my study forever, so he was moved into our old chicken yard, which we had made bird-proof to prevent sparrows stealing all the chicken food. Small settled into his new home, but always greeted David or me with wild wing-waggles and much squealing, and he would run or fly towards us to be fed. We put a mixture of seeds in some shallow dishes, and scattered some around the chicken yard and, before long, he started to peck at it and eat it.

Then, he gave me a terrible fright! I went out one morning and, instead of squealing at me, he croaked! It had been a cool night, and I thought he had caught pneumonia or some horrible throat infection. I could do nothing but wait. He seemed well enough and was still hungry for crop milk. Two anxious days passed – and then, instead of croaking, he said, "Woo! Woo!" His voice had broken! And after that, he was no longer Small but Woo.

When I was reasonably confident that Woo could fend for himself if necessary, I began to let him follow me into our garden and he started to explore it – sometimes by running over the ground, and sometimes by flying around. At first, I would take him back to the chicken yard when it was time for me to go back to my study, but then I started leaving him outside in the garden. He never went far away and, when I appeared again, he would call, "Woo" and I would answer with an imitation of this call. Then he would come running or flying and, as often as not, he ended up on my shoulder or my head. David received similar treatment except that Woo's favourite place was on top of his head and he would often settle down to have a nap there!

In the chicken yard, Woo had a perch under a sheet of corrugated-iron where it was dry even when it rained. For several weeks, I always locked him up in the chicken yard at night, but then I began to leave the door open so he could come and go at will. However, I always went out after dark to make sure he was all right and, one night, he was not there! I knew it was inevitable that he would leave us one day, and this was what we wished for, but I felt sad as I walked back to the house. I need not have worried. As I walked past our lovely large guelder rose bush, a soft "Woo" came from its depths. I called "Woo" back and he answered me – and then I was very happy.

Woo began to fly further afield but he always seemed to spot us if we came into the garden, and he would fly to us immediately. We would see him, as a tiny spot in the sky, and then watch as he became pigeon-like – flying swiftly with his feet tucked tightly into his chest feathers. Then, like an aeroplane coming down to land, his feet would come down like wheels and he would land and topple forward, in typical Crested Pigeon fashion, before righting himself. He never missed coming down out of the blue when we had our lunch in the garden. He would land on the table and help himself to pecks of bread or cheese or anything else he thought worth trying. Then he would fly onto David's head or mine and nestle into our hair for his postprandial nap.

There were other Crested Pigeons who came to be fed at Dossie Lattin's house half-way along Spencer Street, and there was a pair in the garden of the house next to ours along Boldrewood Street, but we never had any in our garden even though we put birdseed out for other birds. However, as Woo flew further afield, it was inevitable that he would encounter other Crested Pigeons and, as luck would have it, I think we observed his first encounter with the pair who lived next door. The encounter took place in a huge liquidambar tree where the resident pair was perching. There was much 'woo-ing' and displaying with head-bobbing, and wing movements and the adopting of a variety of postures. And, after that, Woo stayed with this pair most of the time.

However, if I walked past the garden next door and the pigeons were there, Woo would call "Woo" to me and I would answer, and then he would usually leave the pair and fly onto my shoulder. This was lovely for me, but it distressed the pair and before long, Woo stopped coming to me although he would always call to me and answer if I called to him. Sometimes, I would be walking in the neighbourhood, aware that Crested Pigeons were feeding on the nature strip or perching in nearby trees but not knowing if one of them was Woo. Then one of them would call and I would know it was Woo because other pigeons never called when humans were near, and they never answered my imitation calls as Woo always did.

Woo was still around eight years later. He often came to our bird table where there was always food suitable for pigeons, and he brought other Crested Pigeons to the bird table too. After that, Crested Pigeons came fairly regularly for food and sometimes I would get an answer if I called them. Crested Pigeons still come, and they will often attract my attention to their need for food by calling "Woo! Woo!" I don't know how long Crested Pigeons live but I suspect Woo has long since gone. However, it is not beyond the bounds of possibility that it is some of his offspring who still come, and that they had learned to attract our attention by copying our Woo.

We miss Woo. Of all the birds we ever reared, we think of him (or perhaps he was her) as being the most endearing.

Blossom Possum

At home, in Canberra, we had the bird-proof yard in which we had kept chickens and bantams until we went to Malawi in 1993 and, built on to the chicken yard was a chicken shed in which they could roost at night. Soon after we parted with the chickens and bantams, a wild Brushtail Possum broke into the chicken shed by scratching a hole through a section of the outside wall which was made of chipboard that had started to disintegrate. The hole was at ground level and therefore within reach of foxes, so I blocked it up with solid timber. However, we had no objection to possums spending the day inside the chicken shed, so I cut another access hole higher up, mounted a nest-box in the top left corner of the shed and suspended a plank of wood so a possum could easily climb from the new entrance hole to the nest-box. The nest-box rested on a shelf made of a wide plank of wood, and we could place titbits on this shelf to entice any resident possums out of the box so we could see them.

The chicken shed proved a popular domicile for possums and, over the years, several came and went, and we could occasionally feed them so they became tame. One, in particular, won our hearts and we named this gentle creature Blossom Possum. She lived in the chicken shed for many years and became tame enough to let us look inside her pouch and follow the development of her joeys. She allowed her female joeys to sleep with her in the nest-box until they were fully grown, and they too became very tame. On the other hand, the male joeys were driven off to find new territories and new places in which to spend the day.

Then, one day in 2012, Blossom Possum came home to her nest-box with one of her eyes swollen and filled with pus. We consulted a vet who said nothing could be done to save the eye, and he advised us just to watch and wait and only intervene if the possum appeared to be suffering. She did not appear to be suffering and, fairly soon, the pus disappeared and then the eye atrophied and eventually disappeared. It seemed that Blossom Possum learned to manage well enough despite being blind in one eye, but we decided to make life easier for her by providing a regular midday meal of fruits and a little bread, and she soon expected it. Very often, if we brought the food later than usual, we would find her wide awake, waiting impatiently on the shelf on which the nest-box rested.

Blossom Possum not only won *our* hearts, she was also adored by Maria who lived next door, and Maria was very willing to take over the daily feeding routine when we were away. In fact, she enjoyed her contact with Blossom Possum so much that we were happy for her to continue feeding her even when we were at home.

One morning, Maria came to our house in a very agitated state. "Meredith," she said, "can you come quickly? The possums have got under the net over our apricot tree, and they can't get out again!" I followed Maria next door and there, sure enough, were two very unhappy possums crouched, in the blazing sunlight, at the foot of the trunk of the apricot tree – Blossom Possum and her almost fully-grown daughter. They blinked at us in the bright light, and we could almost read their thoughts. "Please get us out of here! This is *not* a nice place for possums! This miserable tree has very few leaves and its fruit is not ripe enough to eat. We are starving and horribly hot, and we want to go home!"

Maria and I opened a wide gap in the bird net, but the possums were too bewildered to find the way out. So, I fetched the pet-carrier that Lena had sometimes used to transport her Chihuahua, and I put some bread and apple inside it. I crawled under the bird net and placed the open end of the carrier in front of Blossom Possum. She went in immediately, and then I was able to push her hesitating daughter inside as well. Both possums began to eat ravenously, and they never stopped eating for an instant as I picked up the carrier and took them home to the chicken shed!

That pet-carrier proved useful on another occasion. We had a very severe heatwave in Canberra during the summer of 2013–14, and I was worried that Blossom Possum would be getting cooked in the nest-box under the corrugated-iron roof of the chicken shed, even though a large mulberry tree provided dense shade over the shed. For the first two days, I took water to the nest-box, and Blossom Possum drank and drank. She had been licking her arms and hands to cool herself, so she had become rather dehydrated. Then, as the temperature soared into the forties, I decided to coax her into the pet-carrier and bring her into the house where it was much cooler. She did not mind this and subsequently let herself be carried into the cool house every day until the heatwave ended.

We do not know when Blossom Possum was born, but by 2014 she was showing signs of great age. She had one last joey who, of course, rode on her mother's back until she was old enough to move around independently, but it became apparent that Blossom Possum had stopped being able to groom the dense fur on her back where the joey had ridden. It seemed that her hindlegs could no longer reach far enough to comb the matted fur, and perhaps she was also too stiff to reach the matted fur with her front paws or tongue. This was not a problem during the warm summer months but, as winter approached, we became worried that the matted fur would not insulate the possum from the cold at night. So I decided to see if she would let me brush her rump with our dog's brush. Yes, she let me do this, but the brush was too soft to have any effect on the dirty matted fur. So I tried combing the tangle with the metal bristles on the other side of the dog's brush. She put up with this too, although it must have been very unpleasant – but it did not work either. The only other thing I thought I could attempt was to bathe the possum's rump with warm water, and she even allowed me to do this! However, this did not work either and I was left with a very wet possum to contend with on a very cold day. What could I do? Well, I fetched my hair-dryer, attached it to a long extension cord, and then switched it on. The loud roaring noise was too much for poor Blossom Possum! She leapt for the exit hole, but I had anticipated this and I blocked the way out with my hand. The possum gradually settled down and became used to the loud noise and then, when she felt the lovely warm air on her back, she relaxed completely and enjoyed the warmth until her fur was dry again.

Blossom Possum survived the cool nights of autumn but, by the onset of winter in June, her hindlimbs were so arthritic that she was finding it difficult to climb. To make matters worse, Maria and her family, and David and I, were going away for the rest of the winter and there was no one to take care of the dear old possum. David and I did not think it was right to let her suffer a slow, miserable death in the wild, so we took her to a very sympathetic vet who kindly put her to sleep. His words comforted us – "She was ready to go."

We brought her home and laid her to rest in a grave quite near to that of our much-loved dog Moya, and a lovely, water-worn, granite rock marks the site. Blossom Possum was very special and we will never forget how sweet and gentle she was, and how *enormous* was the trust she had in the humans who were her friends.

Elephants can be so careful

In 1999, David and I went to Zimbabwe and we were invited to participate in the annual game count at Mana Pools National Park on the bank of the great Zambezi River. The count took place in the dry season when animals congregated in a narrow strip of flat land near the river, and separate counts were conducted, on foot, twice daily, for two days. The participants camped in a designated but unfenced camping ground close to the river and, because it was very hot, David and I set up our little two-man tent under the shade of a large *Acacia albida* tree. That was not as sensible as it seemed to be at the time. On our second night, we woke up at about one o'clock, not knowing what had disturbed us. Rather cautiously, we peeped out through a gap in the flyscreen door of the tent – and there was an elephant just few metres away! And, within a few minutes, this one was joined by four more! They had come to feed on the seeds of the acacia, and they were getting them by shaking the trunk of the tree and then hoovering up any seedpods that fell to the ground or over our tent. It was absolutely thrilling! It was bright moonlight and we could see the shadows of the elephants on the sides of the tent. I have to admit it was a bit scary too, but we just stayed very still and watched. Also, we could hear what was going on, but one noise mystified us. It was a rasping, whooshing noise as though one of the elephants were dragging a tarpaulin to and fro on the ground. This was plausible because a friend of ours had decided to sleep in the open back of his pickup truck and we knew he had a tarpaulin with him in case he got cold.

The following morning, we could hardly believe our eyes. There were elephant footprints between our tent and the tent pegs! The elephants had actually stepped over the guy ropes without touching a single one of them whereas, in contrast, we were always stumbling over them.

The origin of the mystery noise was solved later on that day. I had opted out of one of the game counts, so I was alone in the campsite. We had erected a small and rather flimsy camp-table under the acacia tree and I was trying to write some notes there, but I was distracted by a huge elephant who had waded out into the river to graze on herbage on an island. I took photos of him and then, after a while, he emerged and came towards the camp and the acacia tree under which I was sitting. I decided to get up and move closer to a toilet-block where I would be safe, but I watched as the elephant strolled up to the tree, walked between the tree and the camp-table, and then began to scratch his flanks on the tree trunk! That was the rasping noise we had heard in the night! It was quite incredible. There was a thermos flask on the table, and the gap between the table and the tree was only 1.2 metres wide, but the elephant was so careful that he managed to avoid bumping the table and even the thermos remained upright.

Rats that hunt for landmines in Africa

In 2003, we went to the 9th Symposium on African Small Mammals, which was held at the Sokoine University of Agriculture in Morogoro – a lovely old town at the base of the Uluguru Mountains in Tanzania. While we were there, we were able to watch something very interesting – the training of Gambian Giant Pouched Rats (*Cricetomys gambianus*) to locate plastic landmines! David had raised some baby Giant Pouched Rats in Ibadan before I knew him, and then we had Chieffy, the one we rescued in Malawi during our first trip there (Chapter 9), but Giant Pouched Rats usually only become really tame if they are handled before their eyes open, and they do not often breed in captivity. However, unlike dogs, these rabbit-sized rats have a very *very* highly-developed sense of smell, they are too light to detonate a mine if they walk over it, they never tire of repetitive tasks and they live for around eight years – so training them to hunt for plastic landmines is well worthwhile. The word is passed around that the university wants pregnant Giant Pouched Rats, and the locals bring them! The rats give birth in captivity and then the young are handled when they are very young so that they become very tame. When they are old enough, they are trained to come when they hear the sound of a mechanical click and then they are rewarded with food. Then they are put into a Y-maze and are rewarded with food whenever they choose to go down the tunnel that smells of explosive – the other tunnel leads only to an empty chamber. After a while, the rats only go to the correct chamber and they are only fed if they respond to the smell of explosive. Finally, they are taken into the field in harnesses attached to suspended wires, and they are trained to run to and fro across a narrow tract of land where a few harmless mines have been buried at sites known only to the trainers. If a rat smells a mine, it becomes excited and begins to dig down to what it now thinks of as a source of food. Then clicks are made to call the rat back to the trainer and it is given a morsel of something to eat as a reward. It is then run over the same site from the opposite direction and, if it digs again over the buried mine, it is rewarded again – more substantially this time. The rats work for long hours for six days each week and, during this time, they are only fed when they locate mines. On Sundays, however, they do not have to work and are given all the food they can eat. Eventually, when the rats are fully trained, a similar procedure is followed in open country where real landmines are thought to have been buried – for example in Mozambique, Angola and Cambodia. If two rats have both dug over one site, it is concluded that a live mine is there, and then an expert is sent in to deal with it. What an amazing achievement! Unlike metallic mines, which can be located with magnetometers, plastic landmines have proven almost impossible to locate without using animals such as dogs, mongooses or bees. But it seems that the most reliable are the Gambian Giant Pouched Rats.

The last demands of Mammals of Africa prior to publication

It was hard to believe. After years of uncertainty and stress, *Mammals of Africa* was going to be published by Bloomsbury. It was the beginning of a mad gallop along a road that was as rocky as ever. We were given a short opportunity to update profiles because new species had been described, some old species had had their names changed, new distribution records had been published and the conservation status of some species had been revised. Then we had to cope with copy-editing and the designing of pages, and finally the correcting of page-proofs of the volumes we were involved with. But the long, hard road through *Mammals of Africa* eventually came to an end. It was published in January 2013 and our life after *MoA* began.

14

EPILOGUE

One day, two very heavy bags were dropped by our front door. Each weighed nearly sixteen kilograms and the contents felt like big books! Yes, it was our complimentary copies of *Mammals of Africa*! We were delighted with the production, and the first year of our life after *Mammals of Africa* was also made extra enjoyable by the kind things people said about it and these made all the hardships worthwhile. For me, particularly touching was that the late Professor Jiří Gaisler, a very well-known bat biologist, gave a seminar on three recently published books, and he said that the *Mammals of Africa* section on the Chiroptera was, "The best book on African bats ever written". Another great thrill for all of us was that the American Library Association awarded *Mammals of Africa* the Dartmouth Medal for 2013 because they judged it to be the best new reference work – on any topic – published during that year!

Three score and ten

In October 2015, I had my seventieth birthday – I had had my three score years and ten! I am hoping for a few more years yet – but not too many.

What a lucky seventy years I have had – a happy childhood, a wonderful marriage and partnership with the man who has always been my very best friend, two loving children and five loving grandchildren, many dear friends, many interesting and exciting travels to Africa and Europe and, of course, so many friendships with special animals! I am so grateful.

These seventy years of my life have been full of changes. One that seems very relevant to my memoirs is the swing away from the 'anti-anthropomorphism' approach to the study of animal behaviour, towards the understanding that animal behaviour and human behaviour reflect evolution and have far more in common than was previously thought. When I first studied animal behaviour in the 1960s, there were two main schools of thought. The Europeans, such as Nobel laureates Konrad Lorenz, Niko Tinbergen and Karl von Frisch attributed *almost all* behaviour to instinct. In contrast, the Americans followed B. F. Skinner – the inventor of the Skinner box – who believed *all* behaviour was learned. It seemed to take many years before people realized that some behaviour is innate (instinctive) and that some is learned, for example by trial and error and conditioning. And, just as there was once a sharp dividing line between innate and learned behaviour, there was also a very sharp line between humans and their behaviour, and animals lower down on the phylogenetic tree and *their* behaviour. People seemed to forget that humans evolved from other animals, and that humans and other animals – especially those that are closely related to us – can share behavioural characteristics in the same way that they share anatomical and physiological characteristics. In those days, behaviour such as making and using tools, self-awareness, personalities, emotions such as jealousy and love, and cognitive thinking were thought to be exclusively human. Consequently, the greatest sin students like me could commit was anthropomorphism – treating animals as if they were human. And serious ethologists always referred to individual study animals by numbers – it was most unprofessional to give them names! How arrogant humans were in those days – and some still are.

Nowadays, ethologists are delving deeply into studies of intelligence, memory, communication, consciousness, self-awareness and cognitive thinking, and it is becoming more and more obvious that we have been greatly underestimating the abilities of other animals. The ability of other animals (including birds) to use tools and solve problems is mind-boggling and it is hard to believe that, prior to the work of that wonderful lady, Dame Jane Goodall, on Chimpanzees in around 1960, it was believed that only *humans* could make and use tools! Jane Goodall's observations of chimpanzees making and using tools inspired that game-changing telegram from her supervisor, the famous anthropologist Louis Leakey – "Now we must redefine tool, redefine Man, or accept chimpanzees as humans." These days, we know, for example, that several species of birds, as well as chimpanzees, use thin sticks to prize grubs out of holes in trees, Palm Cockatoos cut and shape sticks to make drumsticks for tapping on trees to make rhythmic drumming sounds, and Black-breasted Buzzards use small rocks to crack open the shells of Emu eggs! And this is amazing – in northern Australia where humans have been lighting fires for many thousands of years, to reduce fuel in fire-vulnerable habitats and for other reasons, Black Kites and also some other raptors have learned to pick up burning sticks and then use them to light other fires so that they can then prey on the insects and other animals that flee from those fires! And how about this? In Japan, a species of crow learned to drop hard-shelled nuts onto roads so that cars would crush the shells and make the nuts accessible. Then they started putting the nuts on pedestrian crossings while the pedestrians had the green lights facing them, and then they could let the cars crush the nuts while the lights were red, and then go out to eat them in safety when the lights turned green again! I find it impossible to believe that these animals are not using cognitive thinking.

But, are they *consciously aware* that they are using their brains to solve problems? In fact, are they actually *conscious* of anything? There is a phenomenon that has given some answers to this question. We humans normally know that we are conscious of seeing obstacles in our way when we cross a room cluttered with furniture and toys. But many of our responses to visual stimuli are automatic and we often respond without having to think about how we are going to respond. It is a particular part of the brain, the visual cortex, that is involved in our conscious awareness of what we see, and if the brain is damaged so that nerve impulses from the eyes do not reach all the way to the visual cortex, a person will be *unaware* that he can actually see perfectly well! That person will be convinced that he is blind but, nevertheless, he will be able to negotiate obstacles without making any mistakes and he will be able to look at simple shapes painted on cards and then move a hand to outline those shapes. This is known as 'blindsight' and it is not restricted to humans. A monkey that was blindsighted in only one eye was trained in a Skinner box that had four buttons that could light up green (one at a time), and a red button that the monkey was trained to press if none of the green lights lit up after an audible signal was given.

If a green button lit up and the monkey pressed it, she got a reward. If no green button lit up, the monkey was rewarded if she pressed the red button. Then the monkey's 'good' eye was covered so she could only see with the blindsighted eye. And what happened? If a green button lit up, she pressed it (which indicated that she had seen it), but then she immediately pressed the red button (which indicated that she thought that she had *not* seen a green button lighting up). This means that these monkeys, as well as humans, normally have conscious awareness that they can see. In other words, being conscious of something is not limited to humans. I do not know if this has been tested in other mammals.

Self-awareness is another intriguing issue, and this has been investigated by studying the responses of animals to their reflections in mirrors. It has now been demonstrated that humans, chimpanzees and orangutans are aware that they are looking at themselves. For example, they use mirrors to look into their mouths and to look at other parts of their bodies that they cannot see directly. If, while one of these animals is anaesthetized, a red mark is placed on an eyebrow, when the animal wakes up and sees its reflection in a mirror, it will touch the red mark with its hand. Interestingly, gorillas fail these mirror tests but, on the primate phylogenetic tree, they occur lower down than chimpanzees and orangutans.

Yes, there have been enormous advances in our understanding of animal behaviour, and I sometimes regret that I have not followed these advances and participated in these studies – but one cannot do everything!

And now, I think, it is time for me to stop writing about my encounters with animals. I have a Golden Retriever, three wild possums, a gang of Gang-gangs, a small flock of Crested Pigeons and several magpies who are demanding my attention! And my beloved husband, David, who asked me to write down all my 'animal stories', would love a cup of tea.

GLOSSARY

adj. = adjective
cf. = confer, compare with, as opposed to
pl. = plural
q.v. = quod vide, 'which see'

accession book: a book recording the accession numbers (*q.v.*) of specimens in a collection together with information such as the specimen's scientific name, sex, age, date of collection, locality where collected, collector's name, reproductive condition, habitat, measurements and weight.

accession number: the unique identification number given to a specimen when it is added to a collection.

agonistic behaviour: social behaviour that repels individuals. In rodents, includes attacking, pursuing with intent to fight, fighting, threatening and nest-defence.

alluvial: describes soil or other materials deposited by flowing water.

amicable behaviour: social behaviour that attracts individuals together. In rodents, includes making contact for the purpose of investigating an individual's scent, huddling, mounting, mutual grooming.

amphipod: small shrimp-like crustaceans, having no carapace and a laterally compressed body, belonging to the order Amphipoda. Found in saltwater and freshwater habitats and in some very moist terrestrial habitats.

anthropomorphism: in ethology (*q.v.*), the attributing of human characteristics to the behaviour of other animals.

aspect ratio: the width of a wing relative to its length. Calculated by dividing the square of the wingspan by the wing area. Long, narrow wings have high aspect ratios.

bat detector: device for detecting the presence of bats by converting the frequencies of their ultrasonic echolocation calls (*q.v.*) to frequencies that are audible to humans.

billabong: Australian term for an oxbow lake, an isolated lake or lagoon left behind when a river changes its course.

biomass: the total weight of an organism in a specified area or volume.

bipedal: standing on the hindlimbs, or moving without using the forelimbs.

brown fat: in bats, adipose tissue (fat) specifically used for the generation of heat during arousal after torpor (*q.v.*) or hibernation (*q.v.*).

bund: an embankment or raised causeway following a contour, to control the flow of water over land and prevent erosion.

camera lucida: an optical device (such as a modified binocular microscope) that superimposes the image of something under the microscope upon a surface (e.g. paper) on which the viewer can draw. Enables biologists to make very accurate drawings of organisms or anatomical structures observed under the microscope.

carrying capacity: the maximum density of individuals of one or more species that the habitat can support at any particular time.

CF calls: in bats, echolocation calls (*q.v.*) in which the frequency (pitch) remains constant during most of the call, so that a sonogram of the call (*q.v.*) is a straight, horizontal line except perhaps at the beginning and/or end of the call (*cf.* FM calls).

cf.: in taxonomy, precedes the species name if there is uncertainty about it.

chambo: mainly three species of fish in the genus *Oreochromis*, found in Lake Malawi, that were very popular as a source of food, but overfishing has resulted in all three being listed as Critically Endangered species in the IUCN Redlist.

Chichewa: language of the Chewa people, spoken in Malawi and some adjacent countries.

clutter: in bats, refers to the amount of obstacles such as vegetation and buildings that are detected by a bat's echolocation (*q.v.*) in a particular environment. May vary from absent in the open sky high above the ground to very dense in the canopy of a tree, and in vegetation close to the ground.

cognitive thinking: thought processes involving, for example, reasoning, understanding, forming concepts and associations, solving problems and using innovations.

dambo: seasonally marshy grassland.

deforestation: removal of trees from an area.

Devonian: geological period *c*. 359 to 419 million years ago.

dominance: in ethology (*q.v.*), having a higher status in a social hierarchy or pecking order. In ecology, the greater abundance of one or more species of plants (or animals) in a particular community.

Doppler shift: in bats, in context of echolocation (*q.v.*), the change in pitch (frequency) of an echo resulting from the movement of the bat towards or away from the object reflecting the echolocation call, or from the movement of this object towards or away from the bat, or a combination of both.

echolocation call: in bats, the ultrasonic (*q.v.*) sound emitted for the purpose of echolocation (*q.v.*).

echolocation: the use of reflected ultrasonic (*q.v.*) pulses of sound to perceive the surroundings (including obstacles, prey and other animals).

ecotone: a transitional, boundary area between two zones of vegetation (e.g. forest and grassland) containing components of both and/or plants that specialize in transitional zones.

endemic: restricted to, peculiar to, or prevailing in, a specified country or region.

ethogram: a written description or comprehensive inventory of the behaviour of an animal.

ethology: the scientific study of animal behaviour, usually under natural conditions.

eyeshine: light reflected from the tapetum lucidum at the back of the eyes of a nocturnal animal when, for example, a torch or spotlight is shone into its eyes at night.

FM calls: in bats, echolocation calls in which the frequency (pitch) increases or decreases so that a sonogram (*q.v.*) of the call is not a straight, horizontal line (*cf.* CF calls).

foraging strategy: the behaviour adopted by an animal to obtain its food.

forb: herbaceous flowering plant that is not a grass, sedge or rush.

Fulani: an ethnic group of people found in West Africa including some who are nomadic pastoralists (cattle Fulani) and others who have become sedentary in urban areas (town Fulani).

gaff: a stick with a hook or series of hooks for catching fish from the surfaces of lakes, rivers, seas etc.

harp-trap: trap designed to catch bats without harming them. Comprised of one or more banks of vertically strung 'wires' made of fine, nylon fishing line, suspended over a deep trough (made of canvas or similar material) into which bats fall after flying into the banks of 'wires'. Bats are prevented from climbing out of the trough by flaps of overhanging material under which the bats usually settle and rest.

Hausa: the language most commonly spoken by many different ethnic groups of people in sub-Saharan West Africa, including northern Nigeria. Also, people who speak Hausa.

hibernaculum (*pl.* **hibernacula**): a place, domicile or roost where an animal hibernates (*q.v.*).

hibernation: a state of inactivity accompanied by a reduction in metabolic rate (*q.v.*), lower body temperature, and slow breathing. Occurs in some animals when the ambient temperature is low and food is scarce; usually lasts for weeks or months (*cf.* torpor).

home-range: the area routinely used by an animal for its day-to-day activities and requirements, and which contains the resources required for its survival and reproduction (*cf.* territory [*q.v.*]).

innate behaviour: behaviour that is genetically encoded and requires no learning.

inselberg: isolated rocky hill.

intromission: the process of inserting the penis into the vagina during copulation.

isohyet: a line on a map connecting localities having the same amount of rainfall in a given period.

joey: a baby marsupial, especially larger marsupials such as kangaroos, wombats, possums.

K-strategies: lifestyles that result in maximum survival and reproductive success in habitats with stable carrying capacities (*q.v.*) where population densities are consistently high. K-strategies include defending territories (*q.v.*), producing smaller numbers of offspring less often, maturing comparatively slowly and having comparatively long lifespans (*cf.* r-strategies [*q.v.*]).

katundu: baggage (in Chichewa).

khondi: verandah (in Chichewa).

kHz: the unit of measurement of frequencies of sound waves and electromagnetic waves. One kilohertz = 1000 cycles per second.

labile: unstable or liable to change.

laterite: reddish soil, gravel and rock formed by the weathering of igneous rocks in moist, tropical and subtropical climates.

light-tagging: attaching a light (e.g. a pill-capsule filled with fluorescing chemicals) to a bat so that its flight and foraging behaviour can be observed at night.

limnology: the scientific study of the biological, chemical and physical characters of lakes, rivers and other bodies of freshwater.

live-trap: a trap that captures animals without harming them.

macropod: member of the kangaroo family.

manoeuvrability: in bats, refers to the space required by an individual to alter its flight path while flying at a fixed speed.

metabolic rate: the amount of energy used by an animal per unit time.

metabolic water: water produced by oxidative processes within the body; an important source of water for arid-adapted mammals when free (drinking) water and water within the food is in short supply or unavailable.

microbat: originally the common name for bats in the suborder Microchiroptera. Used here as the vernacular name of African bats in all families except fruit bats in the family Pteropodidae.

microhabitat: a small, specialized habitat within a larger habitat.

Miocene: geological epoch (within the Tertiary period), *c*. 5 to 23 million years ago.

miombo: a vernacular name applied to trees in the genus *Brachystegia*; a type of savanna woodland where *Brachystegia* spp. are the commonest trees or one of the commonest trees.

mist-net: rectangular net made of fine netting (usually monofilament, sometimes two-ply), hung vertically between two upright poles. Several taut, horizontal strands of stronger thread ensure that the net hangs in such a way that a series of horizontal pockets are formed. Bats or birds flying into mist-nets usually fall into a pocket and become caught. Mist-nets are usually 2.4 m wide and from 5.5 to at least 12.8 m long. They can be set at ground level or elevated. Mist-nets need very frequent monitoring to prevent injury and stress to captured animals.

mole-rats: rodents belonging to the families Bathyergidae and Spalacidae, which live mostly underground, in burrows, and feed on underground roots, bulbs and rhizomes. They have cylindrical bodies, broad heads, very small or absent ear pinnae, very small eyes, very short or absent tails, stocky limbs with enlarged claws on the forelimbs, and forward-tilting incisors adapted for chiselling and scraping through hard soil and roots.

monoestrous: in bats, defined as having one litter/year. For species and/or populations that are described as monoestrous, each female has one litter/year. *cf.* polyoestrous (*q.v.*).

monogamy: a mating system in which one male mates with only one female and vice versa.

neonate: newly born animal.

noseleaf: fleshy outgrowth on the dorsal surface of the muzzle of bats in some families, associated with the modification of their echolocation calls.

oestrus: the reproductive period during which ovulation occurs and a female is ready for mating.

ontogeny: the development of an organism from fertilization and embryonic growth to maturity.

panga: a type of machete with a hook-shaped tip commonly used in Africa.

pheromone: a chemical produced by an animal that alters the behaviour of animals of the same species. For example, pheromones produced by newly emerged female moths attract males of the same species for the purpose of mating.

phylogenetic tree: a branching diagram showing the evolutionary history and line of descent of a species or higher taxon (*q.v.*).

Pleistocene: geological epoch (within the Quaternary period), *c.* 10,000 to 1.7 million years ago.

Pliocene: geological epoch (within the Tertiary period), *c.* 2 to 5 million years ago.

polygyny: a mating system in which one male mates with several females. *See also* resource-defence polygyny (*q.v.*). *cf.* monogamy (*q.v.*).

polyoestrous: in bats, defined as having two or more litters/year. For species and/or populations that are described as polyoestrous, each female has two or more litters/year. *cf.* monoestrous (*q.v.*).

post-partum: after birth.

primate: an animal belonging to the order Primates that includes, for example, humans, apes, monkeys and galagoes.

r-strategies: lifestyles that result in maximum survival and reproductive success in environments with large, unpredictable fluctuations in carrying capacity (*q.v.*) where population densities rise rapidly and then fall catastrophically. r-strategies promote rapid population growth when carrying capacity increases. They include rapid onset of reproduction, large litter-sizes, rapid development of young, early attainment of maturity and comparatively short lifespans. *cf.* K-strategies (*q.v.*).

raptors: carnivorous birds of prey such as eagles, falcons, hawks, kites and vultures.

refugia (*sing.* refugium): places where animals can survive when conditions elsewhere become unfavourable.

renal: pertaining to the kidneys.

reproductive chronology: the timing and duration of events such as spermatogenesis (*q.v.*), copulation, ovulation, gestation, parturition, lactation and reproductive inactivity, throughout the year.

resource partitioning: dividing the resources of a habitat in such a way that different species can co-exist without excessive competition for resources such as foods, foraging places and domiciles.

resource-defence polygyny: a mating system in which a male controls access to several females indirectly, by monopolizing critical resources (Emlen & Oring 1977).

rhizome: a continuously growing underground horizontal stem that often puts out roots and shoots.

Rhodesian boiler: typically, a forty-four-gallon drum suspended over a log fire with pipes to carry the hot water down to a shower, bath or tap inside a dwelling.

ringer: an Australian stockman.

riparian: pertaining to the banks of rivers or streams.

rondavel: traditional circular African dwelling with a conical thatched roof.

roost-fidelity: in bats, returning to roost at the same place, day after day.

savanna: grassland that may or may not also support some trees and/or shrubs.

sclerophyll forest: a typically Australian type of vegetation having plants such as eucalypts, acacias and banksias, which have hard leaves that resist wilting and reduce loss of water especially during hot, dry summers.

seed-corn: seed that is saved from one year's harvest for the subsequent planting; may refer to hybrid varieties that result in greater productivity.

serendipity: the happening of enjoyable or beneficial events by chance.

Skinner box: a cage with buttons or levers which animals push to get rewards; used to study operant conditioning and some other aspects of animal behaviour. Invented by B. F. Skinner.

social organization: the nature of relationships between individuals of one species that interact with each other either to form groups or to keep themselves apart except when mating or rearing young. Social organizations of groups are described by characteristics such as sexual composition, age structure, mating system, territoriality, co-operative behaviour and responses to animals of the same species that are not group-members.

sonogram: a graph of the frequency (pitch) of a sound (vertical axis) against time (horizontal axis), which provides a visual image of the main characteristics of the sound.

sperm competition: competition between sperm from different males to fertilize an ovum when a female has mated with several males and then stored (*q.v.*) all their sperm until she ovulates.

sperm storage: storage of sperm in the reproductive tract of males for some time before copulation, or in the reproductive tract of females for an extended period before ovulation takes place. A type of reproductive delay that, in females, lengthens the length of the period between copulation and giving birth.

spermatogenesis: the formation of sperm in the testes.

tabular massif: a large, flat-topped mountain mass that is roughly table-shaped.

taxon (*pl.* **taxa**): any defined taxonomic unit (e.g. family, genus, species, subspecies) in the classification of organisms.

taxonomy: the science of biological nomenclature; the study of the rules, principles and practice of naming and classifying species and other taxa (*q.v.*).

territory: an area defended by an individual against certain other members of the species, usually by overt aggression or advertisement; territories are marked by the urine, faeces or glandular secretions of the territory holder. *cf.* home-range (*q.v.*). The boundary of a territory is a line across which the status of the territory holder changes from dominant (*q.v.*) to subordinate.

theriological: pertaining to mammals.

Torpor (*adj.* **torpid**): a state in which there is a lowering of metabolic rate (*q.v.*) and body temperature when the ambient temperature drops. Arousal from torpor normally occurs when the ambient temperature increases and without high energy costs to the individual. Torpor is associated with a state of inactivity and reduced responsiveness to stimuli. Torpor lasts for only short periods of time (hours or days) (*cf.* hibernation).

tung: a tree (*Vernicia fordii*) producing nuts from which tung oil can be extracted. Tung oil is used for finishing or protecting timber. When exposed to air, coats of tung oil are hard, transparent and almost look wet.

ultrasonic: of sounds, having a frequency (pitch) greater than that which can be heard by humans (greater than about 22 kHz).

vascular plants: clubmosses, horsetails, ferns, conifers and flowering plants. These plants have vascular tissues – xylem, which conducts water and solutes from roots to leaves, and phloem, which conducts the products of photosynthesis from the leaves to the rest of the plant.

venation: the pattern of veins in leaves and in the wings of insects and bats.

vertebrates: animals with backbones including mammals, birds, reptiles, amphibians and fish.

visual cortex: that part of the cortex of the brain that plays an important role in the processing of visual information and, in at least some mammals, the part responsible for the conscious awareness of being able to see.

volant: able to fly.

voucher (= **voucher specimen**): specimen of a species that has been studied, which has been lodged in an accessible collection to serve as a reference. Voucher specimens are particularly essential if the identification of the species needs confirmation, or if its taxonomic status has changed or is likely to change.

wing-loading: in bats, the mass of the bat divided by its wing-area.

wing morphology: in bats, the size and shape of a wing, and related characteristics such as aspect ratio (*q.v.*) and wing-loading (*q.v.*) that influence the characteristics of flight, flying ability, foraging behaviour and other aspects of flying.

wingspan: in bats, the distance between the tips of each wing; but this can be measured using several different methods which give different results.

Yoruba: an ethnic group from south-west and north-central Nigeria and southern and central Benin. Also, the language they speak.

Appendix A

SCIENTIFIC NAMES OF ANIMALS
REFERRED TO BY THEIR VERNACULAR NAMES

African Civet, *Civettictis civetta*

African Lungfish, *Protopterus annectens*

African Straw-coloured Fruit Bat, *Eidolon helvum*

African Toad, *Bufo regularis*

Angolan Collared Fruit Bat, *Myonycteris angolensis* (formerly *Lissonycteris angolensis*)

Armyworm, *Neocleptria puncifera*

Australian Lungfish, *Neoceratodus forsteri*

Australian Magpie, *Gymnorhina tibicen*

Banana Pipistrelle, *Pipistrellus nanus*

Banded Mongoose, *Mungos mungo*

Barn Owl, *Tyto alba*

Bilby, *Macrotis lagotis*

Black Mamba, *Dendroaspis polylepis*

Black Rat, *Rattus rattus*

Black Rhino = Black Rhinoceros, *Diceros bicornis*

Blotched Blue-tongued Lizard, *Tiliqua nigrolutea*

Blue Monkey, *Cercopithicus mitis*

Bottlenose Dolphin, probably Indo-Pacific Bottlenose Dolphin, *Tursiops aduncus*

Broad-toothed Rat, *Mastacomys fuscus*

Brolga, *Grus rubicundus*

Brown Antechinus, *Antechinus stuartii*

Brushtail Possum, *Trichosurus vulpecula*

Buffalo (African), *Syncerus caffer*

Bush Rat, *Rattus fuscipes*

Bushbuck, *Tragelaphus scriptus*

Bushpig, *Potomochoerus larvatus*

Cape Turtle Dove, *Streptopelia capicola*

Chief Rat = Gambian Giant Pouched Rat, *Cricetomys gambianus*

Chimpanzee, *Pan troglodytes*

Clam shrimp, *Conchostraca* sp.

Common Bentwing Bat, *Miniopterus schreibersii*

Common Bulbul, *Pycnonotus barbatus*

Common Duiker, *Sylvicapra grimmia*

Common Eland, *Tragelaphus oryx*

Common Wombat, *Vombatus ursinus*

Crawshay's Zebra, *Equus quagga crawshayi*

Crest-tailed Mulgara, *Dasycercus cristicauda*

Crested Pigeon, *Ocyphaps lophotes*

Cusimanse Mongoose, *Crossarchus obscurus*

Dalton's Soft-furred Mouse, *Praomys daltoni*

Desert Mouse, *Pseudomys desertor*

Dingo, *Canis dingo*

Driver Ant, *Dorylus nigricans*

Dusky Antechinus, *Antechinus swainsonii*

Dusky Hopping-mouse, *Notomys fuscus*

Dusky Sengi, *Elephantulus fuscus*

Eastern Brown Snake, *Pseudonaja textilis*

Eastern Grey Kangaroo, *Macropus giganteus*

Echidna = Short-beaked Echidna, *Tachyglossus aculeatus*

Edward's Swamp Rat, *Malacomys edwardsii*

Egyptian Rousette, *Rousettus aegyptiacus*

Egyptian Slit-faced Bat, *Nycteris thebaica*

Elephant = African Elephant, *Loxodonta africana*

Eurasian Magpie, *Pica pica*

Fairy shrimp, *Branchinella* sp.

Fat-tailed Dunnart, *Sminthopsis crassicaudata*

Fawn Hopping-mouse, *Notomys cervinus*

Franquet's Epauletted Fruit Bat, *Epomops franqueti*

Gang-gang, *Callocephalon fimbriatum*

Greater Galago = Large-eared Greater Galago, *Otolemur crassicaudata*

Griselda's Grass Mouse, *Lemniscomys rosalia*

Heath Rat, *Pseudomys shortridgei*

Herring Gull, *Larus argentatus*

Hippo = Hippopotamus, *Hippopotamus amphibius*

Hooded Rat, domesticated *Rattus norvegicus*

Hopping-mice, *Notomys* spp.

Impala, *Aepyceros melampus*

Kemp's Gerbil, *Gerbilliscus kempi*

Koala, *Phascolarctos cinereus*

Kowari, *Dasyuroides byrnei*

Kultarr or Marsupial Jerboa, *Antechinomys laniger*

Kurrichane Thrush, *Turdus libonyana*

Large-footed Myotis = Large-footed Mouse-eared Bat, *Myotis adversus*

Large-spotted Genet, *Genetta maculata*

Leopard, *Panthera pardus*

Lesser Egyptian Jerboa, *Jaculus jaculus*

Lesser Kudu, *Tragelaphus imberbis*

Lesser Long-eared Bat, *Nyctophilus geoffroyi*

Lion, *Panthera leo*

Little Epauletted Fruit Bat, *Epomophorus labiatus*

Little Free-tailed Bat, *Tadarida pumila*

Little Northern Free-tailed Bat, *Mormopterus loriae*

Little Penguin, *Eudyptula minor*

Long-haired Rat, *Rattus villosissimus*

Long-necked Tortoise, *Chelodina longicollis*

Long-nosed Potoroo, *Potorous tridactylus*

Long-tailed Mouse, *Pseudomys higginsi*

Long-tailed Pouched Rat, *Beamys hindei*

Magpie, see Australian Magpie

Marsupial Jerboa = Kultarr, *Antechinomys laniger*

Mesic Four-striped Grass-mouse, *Rhabdomys dilectus* (formerly *Rhabdomys pumilio*)

Mitchell's Hopping-mouse, *Notomys mitchelli*

Murray Cod, *Maccullochella peelii*

Mutable Sun Squirrel, *Heliosciurus mutabilis*

Nyala, *Tragelaphus angasii*

Olive Baboon, *Papio anubis*

Olive-backed Sunbird, *Cyanomitra verticalis*

Patas Monkey, *Erythrocebus patas*

Peak-saddle Horseshoe Bat = Blasius's Horseshoe Bat, *Rhinolophus blasii*

Plains Mouse, *Pseudomys australis*

Platypus, *Ornithorynchus anatinus*

Red-bellied Black Snake, *Pseudechis porphyriacus*

Red-cheeked Cordon Bleu, *Uraeginthus bengalus*

Red-eyed Dove, *Streptopelia semitorquata*

Red-necked Wallaby, *Macropus rufogriseus*

Red Bush Squirrel, *Paraxerus palliatus*

Roan Antelope, *Hippotragus equinus*

Rock Hyrax, *Procavia capensis*

Rusty-bellied Brush-furred Rat, *Lophuromys sikapusi*

Sable Antelope, *Hippotragus niger*

Sandy Inland Mouse, *Pseudomys hermannsbergensis*

Shield shrimp, *Triops australiensis*

Shining Thicket Rat, *Grammomys kuru*

Silky Mouse, *Pseudomys apodemoides*

Silvery Mole-rat, *Heliophobius argenteocinerus*

Slender Tateril, *Taterillus gracilis*

Southern Brown Bandicoot, *Isoodon obesulus*

Southern Lesser Galago, *Galago moholi*

Southern Reedbuck, *Redunca redunca*

Spinifex Hopping-mouse, *Notomys alexis*

Spotted Hyaena, *Crocuta crocuta*

Striped Hyaena, *Hyaena hyaena*

Striped Leaf-nosed Bat, *Macronycteris vittatus*

Stripe-faced Dunnart, *Sminthopsis macroura* (formerly *S. larapinta*)

Sugar Glider, *Petaurus breviceps*

Sulphur-crested Cockatoo, *Cacatua galerita*

Suni, *Neotragus moschatus*

Superb Lyrebird, *Menura novaehollandiae*

Swamp Rat, *Rattus lutreolus*

Swamp Wallaby, *Wallabia bicolor*

Tiger Snake, *Notechis scutatus*

Tiny Pygmy Mouse, *Mus minutoides*

Tuatara, *Sphenodon punctatus*

Tullberg's Soft-furred Mouse, *Praomys tullbergi*

Two-spotted Palm Civet, *Nandinia binotata*

Vervet Monkey, *Chlorocebus pygerythrus*

Walrus, *Odobenus rosmarus*

Warthog, *Phacochoerus aethiopicus*

Waterbuck, *Kobus ellipsiprymnus*

Water-rat, *Hydromys chrysogaster*

Water flea, *Daphnia* sp.

Water-holding Frog, *Cyclorana platycephalus*

Western Gorilla, *Gorilla gorilla*

Western Tree Hyrax, *Dendrohyrax dorsalis*

White Rhino = White Rhinoceros, *Ceratotherium simum*

Wildebeest, *Connochaetes taurinus*

Willie Wagtail, *Rhipidura leucophrys*

Woermann's Long-tongued Fruit Bat, *Megaloglossus woermanni*

Wombat, see Common Wombat

Yellow-footed Antechinus, *Antechinus flavipes*

Yellow Baboon, *Papio cynocephalus*

Zebra Grass Mouse, *Lemniscomys zebra*

Zenkers's Fruit Bat, *Scotonycteris zenkeri*

Appendix B

VERNACULAR NAMES OF STUDY ANIMALS REFERRED TO BY THEIR SCIENTIFIC NAMES

Acomys spinosissimus, Least Spiny Mouse

Aethomys chrysophilus, Red Veld Rat

Aethomys namaquensis, Namaqua Veld Rat

Beamys hindei, Long-tailed Pouched Rat

Dasymys incomptus, Common Shaggy Rat

Dendromys nyikae, Nyika African Climbing Mouse

Eidolon helvum, African Straw-coloured Fruit Bat

Epomophorus labiatus, Little Epauletted Fruit Bat

Epomophorus wahlbergi, Wahlberg's Epauletted Fruit Bat

Eptesicus hottentotus, Long-tailed Serotine

Gerbilliscus leucogaster, Bushveld Gerbil

Glauconycteris variegata, Variegated Butterfly Bat

Grammomys dolichurus, Woodland Thicket Rat

Grammomys ibeanus, East African Thicket Rat

Hipposideros ruber, Noack's Leaf-nosed Bat

Kerivoula argentata, Damara Woolly Bat

Laephotis botswanae, Botswanan Long-eared Bat

Lophuromys flavopunctatus, Yellow-spotted Brush-furred Rat

Macronycteris vittatus, Striped Leaf-nosed Bat

Mastomys natalensis, Natal Multimammate Mouse

Miniopterus sp., a long-fingered bat of uncertain species

Mus triton, Grey-bellied Pygmy Mouse

Myotis bocagii, Rufous Myotis

Myotis welwitschii, Welwitsch's Myotis

Nycteris grandis, Large Slit-faced Bat

Nycteris hispida, Hairy Slit-faced Bat

Nycteris macrotis, Large-eared Slit-faced Bat

Nycteris thebaica, Egyptian Slit-faced Bat

Nycteris woodi, Wood's Slit-faced Bat

Nycticeinops schlieffeni, Schlieffen's Twilight Bat

Otomys angoniensis, Angoni Vlei Rat

Pelomys fallax, East African Creek Rat

Pipistrellus hesperidus, Dusk Pipistrelle

Pipistrellus nanus, Banana Pipistrelle (formerly Banana Bat)

Pipistrellus rendalli, Rendall's Pipistrelle

Pipistrellus rueppellii, Rüppell's Pipistrelle

Pipistrellus stanleyi, Stanley's Pipistrelle (formerly cf. *melckorum*, Melcks' Pipistrelle)

Pipistrellus zuluensis, Zulu Pipistrelle

Praomys delectorum, Delicate Soft-furred Mouse

Rattus rattus, Black Rat

Rhabdomys delectus, Mesic Four-striped Grass-mouse (formerly *Rhabdomys pumilio*)

Rhinolophus blasii, Peak-saddle Horseshoe Bat or Blasius's Horseshoe Bat

Rhinolophus clivosus, Geoffroy's Horseshoe Bat

Rhinolophus fumigatus, Rüppell's Horseshoe Bat

Rhinolophus simulator, Bushveld Horseshoe Bat

Rhinolophus swinnyi, Swinny's Horseshoe Bat

Rousettus aegyptiacus, Egyptian Rousette

Saccostomus campestris, Cape Pouched Mouse

Scotoecus albofuscus woodi, Wood's Light-winged Lesser House Bat

Scotoecus hirundo, Dark-winged Lesser House Bat

Scotophilus dinganii, Yellow-bellied House Bat

Scotophilus nigrita, Giant House Bat

Steatomys pratensis, Fat Mouse

Tadarida condylura, Angolan Free-tailed Bat

Tadarida nigeriae, Nigerian Free-tailed Bat

Tadarida pumila, Little Free-tailed Bat

Taphozous mauritianus, Mauritian Sheath-tailed Bat
Triaenops afra, African Trident Bat
Uranomys ruddi (originally *Uranomys woodi*), Rudd's Brush-furred Mouse

FURTHER READING

Aldridge, H. D. J. N. & Rautenbach, I. L. 1987. Morphology, echolocation and resource partitioning in insectivorous bats. *Journal of Animal Ecology* 56: 763–778.

Baudinette, R. V. 1972. The impact of social aggregation on the respiratory physiology of Australian hopping-mice. *Comparative Biochemistry and Physiology* 41: 35–38.

Bernard, R. T. F., Happold, D. C. D. & Happold, M. 1997. Sperm storage in the Banana Bat (*Pipistrellus nanus*) from tropical latitudes in Africa. *Journal of Zoology, London* 241: 161–174.

Emlen, S. T. & Oring, L. W. 1977. Ecology, sexual selection, and the evolution of mating systems. *Science* 197: 215–223.

Happold, David. 2011. *African Naturalist. The Life and Times of Rodney Carrington Wood, 1889–1962*. Book Guild Publishing, Sussex. England. 290 pp.

Happold, D. C. D. 1967. Biology of the jerboa, *Jaculus jaculus butleri* (Rodentia, Dipodidae), in the Sudan. *Journal of Zoology, London* 151: 257–275.

Happold, D. C. D. 1987. *The Mammals of Nigeria*. Oxford University Press, Oxford. 402 pp.

Happold, D. C. D. (ed). 2013. *Mammals of Africa*. Volume III: Rodents, Hares and Rabbits. Bloomsbury Publishing, London. 784 pp.

Happold, D. C. D. & Happold, M. 1978. Fruit bats of western Nigeria. *Nigerian Field* 43: 30–37, 72–77, 121–127.

Happold, D. C. D. & Happold, M. 1985. The natural history of bats in Malawi. *Nyala* 11: 57–62.

Happold D. C. D. & Happold M. 1986. Small mammals of Zomba Plateau, Malawi, as assessed by their presence in pellets of the grass owl, *Tyto capensis*, and by live-trapping. *African Journal of Ecology* 24: 77–87.

Happold, D. C. D. & Happold, M. 1987. Small mammals in pine plantations and natural habitats on Zomba Plateau, Malawi. *Journal of Applied Ecology* 24: 353–367.

Happold, D. C. D. & Happold, M. 1988. Renal form and function in relation to the ecology of bats (Chiroptera) from Malawi, central Africa. *Journal of Zoology, London* 215: 629–655.

Happold, D. C. D. & Happold, M. 1989. Biogeography of montane small mammals in Malawi, central Africa. *Journal of Biogeography* 16: 353–367.

Happold, D. C. D & Happold, M. 1989. Demography and habitat preferences of small mammals on Zomba Plateau, Malawi. *Journal of Zoology, London* 219: 581–605.

Happold, D. C. D. & Happold, M. 1989. The reproduction of *Tadarida condylura* and *Tadarida pumila* (Chiroptera, Molossidae) in Malawi, Central Africa. *Journal of Reproduction and Fertility* 85: 133–149.

Happold, D. C. D. & Happold, M. 1990. Demography and habitat preference of small mammals in Liwonde National Park, Malawi. *Journal of Zoology, London* 221: 219–235.

Happold, D. C. D. & Happold, M. 1990. The domiciles, reproduction, social organization and sex ratios of *Pipistrellus nanus* (Chiroptera, Vespertilionidae) in Malawi. *Zeitschrift für Saugetierkunde* 55: 145–160.

Happold, D. C. D. & Happold, M. 1991. An ecological study of small rodents in the thicket-clump savanna of Lengwe National Park, Malawi. *Journal of Zoology, London* 223: 527–547.

Happold, D. C. D & Happold, M. 1992. The ecology of three communities of small mammals at different altitudes in Malawi, Central Africa. *Journal of Zoology, London* 228: 81–101.

Happold, D. C. D. & Happold, M. 1992. Termites as food for the Thick-tailed Bushbaby (*Otolemur crassicaudata*) in Malawi. *Folia Primatologia* 58:118–120.

Happold, D. C. D. & Happold, M. 1991. The reproductive strategies of bats (Chiroptera) in Africa. *Journal of Zoology, London* 222: 557–583.

Happold, D. C. D. & Happold, M. 1996. The social organization and population dynamics of leaf-roosting banana bats, *Pipistrellus nanus* (Chiroptera, Vespertilionidae), in Malawi, east-central Africa. *Mammalia* 60: 517–544

Happold, D. C. D. & Happold, M. 1997. Conservation of mammals on a tobacco estate on the Highlands of Malawi. *Biodiversity and Conservation* 6: 837–852.

Happold, D. C. D. & Happold, M. 1998. Effects of bat-bands and banding on a population of *Pipistrellus nanus* (Chiroptera: Vespertilionidae) in Malawi. *Zeitschrift für Saugetierkunde* 63: 65–78.

Happold D. C. D., Happold M. & Hill, J. E. 1987. The bats of Malawi. *Mammalia* 51: 337–414.

Happold, M. 1972. Maternal and juvenile behaviour in the marsupial jerboa, *Antechinomys spenceri* (Dasyuridae). *Australian Mammalogy* 1: 27–37.

Happold, M. 1976. Reproductive biology and developments in the conilurine rodents (Muridae) of Australia. *Australian Journal of Zoology* 24: 19–26.

Happold, M. 1976. Social behaviour of the conilurine rodents (Muridae) of Australia. *Zeitschrift für Tierpsychologie* 40: 113–182.

Happold. M. 2005. A new species of *Myotis* (Chiroptera: Vespertilionidae) from central Africa. *Acta Chiropterologica* 7: 9–21.

Happold. M. 2013. Order Chiroptera and other profiles in: *Mammals of Africa* Volume IV (eds. M. Happold & D. C. D. Happold). Bloomsbury, London.

Happold, M. & Happold D. C. D. 1997. New records of bats (Chiroptera: Mammalia) from Malawi, east-central Africa, with an assessment of their status and conservation. *Journal of Natural History* 31: 805–836.

Happold, M. & Happold, D. C. D. (1997). Chiromo and Thyolo revisited: comments on the conservation of small mammals in Malawi. *Nyala* 20: 1–10.

Happold, M. & Happold, D. C. D. (eds). 2013. *Mammals of Africa*. Volume IV: Hedgehogs, Shrews and Bats. Bloomsbury Publishing, London. 800 pp.

Lee, A. K., Fleming, M. R. & Happold, M. 1984. Microclimate, water economy and energetics of a desert rodent, *Notomys alexis*. In: *Arid Australia* (eds. H. G. Cogger & E. E. Cameron). Australian Museum, Sydney. pp. 315–326.

MacMillen, R. E. & Lee, A. K. 1967. Australian desert mice: independence of exogenous water. *Science* 158: 383–385.

MacMillen, R. E. & Lee, A. K. 1970. Energy metabolism and pulmocutaneous water loss of Australian hopping-mice. *Comparative Biochemistry and Physiology* 35: 355–369.

Meadows, M. E. 1984. Late Quaternary vegetation history of the Nyika Plateau, Malawi. *Journal of Biogeography* 11: 209–222.

Morton, S. R., Happold, M., Lee, A. K. & MacMillen, R. E. 1977. The diet of the Barn owl, *Tyto alba*, in South-western Queensland. *Australian Wildlife Research* 4: 91–97.

Stanley, M. 1971. An ethogram of the hopping-mouse, *Notomys alexis*. *Zeitschrift für Tierpsychologie* 29: 225–258.